Lecture Notes in Computer Science

Commenced Publication in 1973
Founding and Former Series Editors:
Gerhard Goos, Juris Hartmanis, and Jan van Leeuwen

Christophe Jermann   Arnold Neumaier
Djamila Sam (Eds.)

Global Optimization and Constraint Satisfaction

Second International Workshop, COCOS 2003
Lausanne, Switzerland, November 18-21, 2003
Revised Selected Papers

 Springer

Volume Editors

Christophe Jermann
Université de Nantes, LINA
BP 92208, 2 rue de la Houssinière, 44322 Nantes, France
E-mail: christophe.jermann@univ-nantes.fr

Arnold Neumaier
University Wien, Institute for Mathematic
Nordbergstr. 15, A-1090 Wien, Austria
E-mail: Arnold.Neumaier@univie.ac.at

Djamila Sam
Swiss Federal Institute of Technology
Artificial Intelligence Laboratory
Route J.-D. Colladon, Bat. INR, Office 235, CH-1015 Lausanne, Switzerland
E-mail: jamila.sam@epfl.ch

Library of Congress Control Number: 2005926499

CR Subject Classification (1998): G.1.6, G.1, F.4.1, I.1

ISSN   0302-9743
ISBN-10   3-540-26003-X Springer Berlin Heidelberg New York
ISBN-13   978-3-540-26003-5 Springer Berlin Heidelberg New York

Springer is a part of Springer Science+Business Media

springeronline.com

© Springer-Verlag Berlin Heidelberg 2005
Printed in Germany

Typesetting: Camera-ready by author, data conversion by Scientific Publishing Services, Chennai, India
Printed on acid-free paper   SPIN: 11425076   06/3142   5 4 3 2 1 0

Preface

The formulation of many practical problems naturally involves constraints on the variables entering the mathematical model of a real-life situation to be analyzed. It is of great interest to find the possible scenarios satisfying all constraints, and, if there are many of them, either to find the best solution, or to obtain a compact, explicit representation of the whole feasible set.

The 2nd Workshop on Global Constrained Optimization and Constraint Satisfaction, COCOS 2003, which took place during November 18–21, 2003 in Lausanne, Switzerland, was dedicated to theoretical, algorithmic, and application oriented advances in answering these questions. Here global optimization refers to finding the absolutely best feasible point, while constraint satisfaction refers to finding all possible feasible points. As in COCOS 2002, the first such workshop (see the proceeedings [1]), the emphasis was on complete solving techniques for problems involving continuous variables that provide all solutions with full rigor, and on applications which, however, were allowed to have relaxed standards of rigor.

The participants used the opportunity to meet experts from global optimization, mathematical programming, constraint programming, and applications, and to present and discuss ongoing work and new directions in the field. Four invited lectures and 20 contributed talks were presented at the workshop. The invited lectures were given by John Hooker (Logic-Based Methods for Global Optimization), Jean-Pierre Merlet (Usual and Unusual Applications of Interval Analysis), Hermann Schichl (The COCONUT Optimization Environment), and Jorge Moré (Global Optimization Computational Servers).

This volume contains the text of Hooker's invited lecture and of 12 contributed talks. Copies of the slides for most presentations can be found at [2].

Constraint satisfaction problems. Three papers focus on algorithmic aspects of constraint satisfaction problems.

The paper *Efficient Pruning Technique Based on Linear Relaxations* by Lebbah, Michel and Rueher describes a very successful combination of constraint propagation, linear programming techniques and safe rounding procedures to obtain an efficient global solver for nonlinear systems of equations and inequalities with isolated solutions only, providing mathematically guaranteed performance.

The paper *Inter-block Backtracking: Exploiting the Structure in Continuous CSPs* by Jermann, Neveu and Trombettoni shows how the sparsity structure often present in constraint satisfaction problems can be exploited to some extent by decomposing the full problem into a number of subsystems. By judiciously distributing the work into (a) searching solutions for individual subsystems and (b) combining solutions of the subsystems, one can often gain speed, sometimes orders of magnitude.

The paper *Accelerating Consistency Techniques for Parameter Estimation of Exponential Sums* by Garloff, Granvilliers and Smith discusses constraint satisfaction techniques for the estimation of parameters in time series modeled as exponential sums, given uncertainty intervals for measured time series.

Global optimization. Five papers deal with improvements in global optimization methods.

The paper *Convex Programming Methods for Global Optimization* by Hooker describes how to reduce global optimization problems to convex nonlinear programming in case the problem becomes convex when selected discrete variables are fixed. The techniques discussed include disjunctive programming with convex relaxations, logic-based outer approximation, logic-based Benders decomposition, and branch-and-bound using convex quasi-relaxations.

The paper *A Method for Global Optimization of Large Systems of Quadratic Constraints* by Lamba, Dietz, Johnson and Boddy presents a new algorithm for the global optimization of quadratically constrained quadratic programs, which is shown to be efficient for large problems arising in the scheduling of refineries, involving many thousands of variables and constraints.

The paper *A Comparison of Methods for the Computation of Affine Lower Bound Functions for Polynomials* by Garloff and Smith shows how to exploit Bernstein expansions to find efficient rigorous affine lower bounds for multivariate polynomials, needed in global optimization algorithms.

The paper *Using a Cooperative Solving Approach to Global Optimization Problems* by Kleymenov and Semenov presents SIBCALC, a cooperative solver for global optimization problems.

The paper *Global Optimization of Convex Multiplicative Programs by Duality Theory* by Oliveira and Ferreira shows how to use outer approximation together with branch and bound to minimize a product of positive convex functions subject to convex constraints. This arises naturally in convex multiobjective programming.

Applications. The paper *High-Fidelity Models in Global Optimization* by Peri and Campana applies global optimization to large problems in ship design. An important ingredient of their methodology is the ability to use models of different fidelity, so that the most expensive computations on high-fidelity models need to be done with lowest frequency.

The paper *Incremental Construction of the Robot's Environmental Map Using Interval Analysis* by Drocourt, Delahoche, Brassart and Cauchois uses constraint propagation based algorithms for building maps of the environment of a moving robot.

The paper *Nonlinear Predictive Control Using Constraints Satisfaction* by Lydoire and Poignet discusses the design of nonlinear model predictive controllers satisfying given constraints, using constraint satisfaction techniques.

The paper *Gas Turbine Model-Based Robust Fault Detection Using a Forward-Backward Test* by Stancu, Puig and Quevedo presents a new, constraint propagation based method for fault detection in nonlinear, discrete dynamical systems with parameter uncertainties which avoids the wrapping effect that spoils most computations involving dynamical systems.

The paper *Benchmarking on Approaches to Interval Observation Applied to Robust Fault Detection* by Stancu, Puig, Cugueró and Quevedo applies interval techniques to the uncertainty analysis in model-based fault detection.

This volume of contributions to global optimization and constraint satisfaction thus reflects the trend both towards more powerful algorithms that allow us to tackle larger and larger problems, and towards more-demanding real-life applications.

January 2005

Christophe Jermann
Arnold Neumaier
Djamila Sam

References

1. Ch. Bliek, Ch. Jermann and A. Neumaier (eds.), Global Optimization and Constraint Satisfaction, Lecture Notes in Computer Science 2861, Springer, Berlin, Heidelberg, New York, 2003.
2. COCOS 2003 – Global Constrained Optimization and Constraint Satisfaction, Web site (2003), http://liawww.epfl.ch/Events/Cocos03

Organization

The COCOS 2003 workshop was organized by the partners of the COCONUT project (IST-2000-26063) with financial support from the European Commission and the Swiss Federal Education and Science Office (OFES).

Programme Committee

Frédéric Benhamou   Université de Nantes, France
Christian Bliek   ILOG, France
Boi Faltings   Ecole Polytechnique Fédérale de Lausanne, Switzerland

Arnold Neumaier   University of Vienna, Austria
Peter Spellucci   Darmstadt University, Germany
Pascal Van Hentenryck   Brown University, USA
Luis N. Vicente   University of Coimbra, Portugal

Referees

C. Avelino   L. Jaulin   N. Sahinidis
F. Benhamou   R.B. Kearfott   J. Soares
C. Bliek   A. Neumaier   P. Spellucci
B. Faltings   B. Pajot   L. Vicente
L. Granvilliers   B. Raphael
C. Jansson   S. Ratschan

Table of Contents

Constraint Satisfaction

Global Optimization

Applications

Efficient Pruning Technique Based on Linear Relaxations

Yahia Lebbah[1,2], Claude Michel[1], and Michel Rueher[1]

[1] COPRIN (I3S/CNRS - INRIA),
Université de Nice–Sophia Antipolis,
930, route des Colles, B.P. 145,
06903 Sophia Antipolis Cedex, France
{cpjm, rueher}@essi.fr
[2] Université d'Oran Es-Senia, Faculté des Sciences,
Département Informatique,
B.P. 1524 El-M'Naouar, Oran, Algeria
ylebbah@sophia.inria.fr

Abstract. This paper extends the `Quad`-filtering algorithm for handling general nonlinear systems. This extended algorithm is based on the RLT (Reformulation-Linearization Technique) schema. In the reformulation phase, tight convex and concave approximations of nonlinear terms are generated, that's to say for bilinear terms, product of variables, power and univariate terms. New variables are introduced to linearize the initial constraint system. A linear programming solver is called to prune the domains. A combination of this filtering technique with `Box`-consistency filtering algorithm has been investigated. Experimental results on difficult problems show that a solver based on this combination outperforms classical CSP solvers.

1   Introduction

Numerical constraint systems are widely used to model problems in numerous application areas ranging from robotics to chemistry. Solvers of nonlinear constraint systems over the real numbers are based upon partial consistencies and searching techniques.

The drawback of classical local consistencies (e.g. 2B-consistency [13] and `Box`-consistency [3]) comes from the fact that the constraints are handled independently and in a blind way. 3B-consistency [13] and kB-consistency [13] are partial consistencies that can achieve a better pruning since they are "less local" [10]. However, they require numerous splitting steps to find the solutions of a system of nonlinear constraints; so, they may become rather slow.

For instance, classical local consistencies do not exploit the semantic of quadratic terms; that's to say, these approaches do not take advantage of the very specific semantic of quadratic constraints to reduce the domains of the variables. Linear programming techniques [1, 25, 2] do capture most of the semantic of quadratic terms (e.g., convex and concave envelopes of these particular terms).

C. Jermann et al. (Eds.): COCOS 2003, LNCS 3478, pp. 1–14, 2005.
© Springer-Verlag Berlin Heidelberg 2005

That's why we have introduced in [11] a global filtering algorithm (named `Quad`) for handling systems of quadratic equations and inequalities over the real numbers. The `Quad`-algorithm computes convex and concave envelopes of bilinear terms xy as well as concave envelopes and convex underestimations for square terms x^2.

In this paper, we extend the `Quad`-framework for tackling general nonlinear system. More precisely, since every nonlinear term can be rewritten as sums of products of univariate terms, we introduce relaxations for handling the following terms:

- power term x^n
- product of variables $x_1 x_2 ..x_n$
- univariate term $f(x)$

The `Quad`-algorithm is used as a global filtering algorithm in a branch and prune approach [29]. Branch and prune is a search-tree algorithm where filtering techniques are applied at each node. `Quad`-algorithm uses Box-consistency and 2B-consistency filtering algorithms. In addition, linear and nonlinear relaxations of non-convex constraints are used for range reduction in the branch-and-reduce algorithm [19]. More precisely, the `Quad`-algorithm works on the relaxations of the nonlinear terms of the constraint system whereas `Box`-consistency algorithm works on the initial constraint system.

Yamamura et. al. [31] have first used the simplex algorithm on quasi-linear equations for excluding interval vectors (boxes) containing no solution. They replace each nonlinear term by a new variable but they do not take into account the semantic of nonlinear terms[1]. Thus, their approach is rather inefficient for systems with many nonlinear terms.

The paper is organised as follows. Notations and classical consistencies are introduced in section 2. Section 3 introduces and extends the `Quad` pruning algorithm. Experimental results are reported in section 4 whereas related works are discussed in section 5.

2   Notation and Basics on Classical Continuous Consistencies

This paper focuses on CSPs where the domains are intervals and the constraints are continuous. A n-ary continuous constraint $C_j(x_1, \ldots, x_n)$ is a relation over the reals. $\mathcal{C}$ stands for the set of constraints.

D_x denotes the domain of variable x, that's to say, the interval $[\underline{x}, \overline{x}]$ of allowed values for x. $\mathcal{D}$ stands for the set of domains of all the variables of the considered constraint system.

We use the "reformulation-linearization technique" notations introduced in [25, 2] with some modifications. Let E be some nonlinear expression, $[E]_L$ denotes the set of linear terms coming from a linearization process of E.

---

[1] They introduce only some weak approximation for convex and monotone functions.

We also use two local consistencies derived from Arc-consistency [14]: 2B-consistency and Box-consistency.

2B-consistency [13] states a local property on the bounds of the domains of a variable at a single constraint level. Roughly speaking, a constraint c is 2B-consistent if, for any variable x, there exists values in the domains of all other variables which satisfy c when x is fixed to $\underline{x}$ or $\overline{x}$.

Box-consistency [3] is a coarser relaxation of Arc-consistency than 2B-consistency. It mainly consists of replacing every existentially quantified variables but one with its interval in the definition of 2B-consistency. Box-consistency [3] is the most successful adaptation of arc-consistency [14] to constraints over the real numbers. Furthermore, the narrowing operator for the Box-consistency has been extended [29] to prove the unicity of a solution in some cases.

The success of 2B-consistency and Box-consistency depends on the precision of enforcing local consistency of each constraint on each variable lower and upper bounds. Thus they are very local and do not exploit any specific semantic of the constraints.

3B-consistency and kB-consistency are partial consistencies that can achieve a better pruning since they are "less local" [10]. However, they require numerous splitting steps to find the solutions of a system of nonlinear constraints; so, they may become rather slow.

3   Using Linear Relaxations to Prune the Domains

In this section, we introduce the filtering procedure we propose for handling general constraints. The Quad filtering algorithm (see Algorithm 1.1) consists of three main steps: reformulation, linearization and pruning.

The reformulation step generates $[\mathcal{C}]_R$, the set of implied linear constraints. More precisely, $[\mathcal{C}]_R$ contains linear inequalities that approximate the semantic of nonlinear terms of $[\mathcal{C}]$.

The linearization process first decomposes each non linear term E in sums and products of univariate terms. Then, it replaces nonlinear terms with their associated new variables. For example, consider $E = \{x_2 x_3 x_4^2 (x_6 + x_7) + \sin(x_1)(x_2 x_6 - x_3) = 0\}$, a simple linearization transformation may yield the following sets:

- $[E]_L = \{y_1 + y_3 = 0, y_2 = x_6 + x_7, y_4 = y_5 - x_3\}$
- $[E]_{LI} = \{y_1 = x_2 x_3 x_4^2 y_2, y_3 = \sin(x_1) y_4, y_5 = x_2 x_6\}$.

$[E]_{LI}$ denotes the set of equalities that keep the link between the new variables and the nonlinear terms.

Finally, the linearization step computes the set of final linear inequalities and equalities $LR = [\mathcal{C}]_L \cup [\mathcal{C}]_R$, the linear relaxation of the original constraints $\mathcal{C}$.

The pruning step is just a fixed point algorithm that calls a linear programming solver iteratively to reduce the upper and the lower bound of each initial variable. The algorithm terminates when the maximum achieved reduction is smaller than a non-null predetermined threshold .

Function Quad_filtering(IN: $\mathcal{X}$, $\mathcal{D}$, $\mathcal{C}$, ϵ) **return** $\mathcal{D}'$
% $\mathcal{X}$: initial variables ; $\mathcal{D}$: input domains; $\mathcal{C}$: constraints; ϵ: minimal reduction, $\mathcal{D}'$: output domains

1. *Reformulation*: generation of linear inequalities $[\mathcal{C}]_R$ for the nonlinear terms in $\mathcal{C}$.
2. *Linearization*: linearization of the whole system $[\mathcal{C}]_L$.
   We obtain a linear system $LR = [\mathcal{C}]_L \cup [\mathcal{C}]_R$.
3. $\mathcal{D}' := \mathcal{D}$
4. *Pruning* :
   While the reduction amount of some bound is greater than ϵ **and** $\emptyset \notin \mathcal{D}'$ **Do**
   (a) Update the coefficients of the linearizations $[\mathcal{C}]_R$ according to the domain $\mathcal{D}'$
   (b) Reduce the lower and upper bounds $\underline{D'_i}$ and $\overline{D'_i}$ of each *initial* variable $x_i \in \mathcal{X}$ by computing *min* and *max* of x_i subject to LR with a linear programming solver.

Algorithm 1.1. The Quad algorithm

Now, we are in position to introduce the reformulation of nonlinear terms. Section 3.1 recalls the relaxations for the simplest case of bilinear term xy, the product of two distinct variables. Relaxations for the power term are given in section 3.2. The process for approximating general product terms is given in section 3.3. Finally, in section 3.4, we introduce a procedure to relax some univariate terms.

3.1 Bilinear Terms

In the case of bilinear terms xy, Al-Khayal and Falk [1] showed that convex and concave envelopes of xy over the box $[\underline{x},\overline{x}] \times [\underline{y},\overline{y}]$ can be approximated by the following relations:

$$[\text{xy}]_R = \begin{cases} \text{B IL}1 \equiv [(x - \underline{x})(y - \underline{y}) \geq 0]_L \\ \text{B IL}2 \equiv [(x - \underline{x})(\overline{y} - y) \geq 0]_L \\ \text{B IL}3 \equiv [(\overline{x} - x)(y - \underline{y}) \geq 0]_L \\ \text{B IL}4 \equiv [(\overline{x} - x)(\overline{y} - y) \geq 0]_L \end{cases} \quad (1)$$

BIL1 and BIL3 define a convex envelope of xy whereas BIL2 and BIL4 define a concave envelope of xy over the box $[\underline{x},\overline{x}] \times [\underline{y},\overline{y}]$. Thus, these relaxations are the optimal convex/concave outer-estimations of xy.

3.2 Power Terms

First let us consider square terms. The term x^2 with $\underline{x} \leq x \leq \overline{x}$ is approximated by the following relations:

$$\text{L1( )} \equiv [(x - \ )^2 \geq 0]_L \text{ where } \ \in [\underline{x},\overline{x}] \quad (2)$$

$$\text{L2} \equiv [(\underline{x} + \overline{x})x - y - \underline{x}\overline{x} \geq 0]_L \quad (3)$$

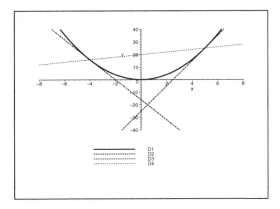

Fig. 1. Approximation of x^2

Note that $[(x -)^2 = 0]_L$ generates the tangent line to the curve $y = x^2$ at the point $x = $. Actually, **Quad** computes only $L1(\overline{x})$ and $L1(\underline{x})$. Consider for instance the quadratic term x^2 with $x \in [-4,5]$. Figure 1 displays the initial curve (i.e., D_1), and the lines corresponding to the equations generated by the relaxations: D_2 for $L1(-4) \equiv y + 8x + 16 \geq 0$, D_3 for $L1(5) \equiv y - 10x + 25 \geq 0$, and D_4 for $L2 \equiv -y + x + 20 \geq 0$.

We may note that $L1(\overline{x})$ and $L1(\underline{x})$ are underestimations of x^2 whereas $L2$ is an overestimation. $L2$ is also the concave envelope, which means that it is the optimal concave overestimation.

More generally, a power term of the form x^n can be approximated by $n + 1$ inequalities with a procedure proposed by Sherali and Tuncbilek [27], called "bound-factor product RLT constraints". It is defined by the following formula:

$$[x^n]_R = \{[(x - \underline{x})^i(\overline{x} - x)^{n-i} \geq 0]_L, i = 0..n\} \qquad (4)$$

The essential observation is that this relaxation generates tight relations between variables on their upper and lower bounds. More precisely, suppose that some original variable takes a value equal to either of its bounds. Then, all the corresponding new RLT linearization variables that involve this original variable take a relative value that conform with actually fixing this original variable at each of its particular bound in the nonlinear expressions represented by these new RLT variables [27].

Note that relaxations (4) of the power term x^n are expressed with x^i for all $i \leq n$, and thus provide a fruitful relationship on problems containing many power terms involving the same variable.

The univariate term x^n is convex when n is even, or when n is odd and the value of x is negative; it is concave when n is odd and the value of x is positive. Section 3.4 details the process for handling such convex and concave univariate term. Sahinidis and Twarmalani [21] have introduced the convex and concave envelopes when n is odd by taking the point where the power term x^n and its under-estimator have the same slope. These convex/concave relaxations on x^n

are expressed with only $[x^n]_L$ and x. In other words, they do not generate any relations with x^i for $1 < i < n$. That's why we suggest to implement these formulas (4).

Note that for the case $n = 2$, (4) provides the concave envelope.

3.3   Product Terms

For the product term

$$x_1 x_2 .. x_n \tag{5}$$

we use a two steps procedure: quadrification and bilinear relaxations.

The *Quadrification* step brings back the multi-linear term into a set of quadratic terms as follows

$$
\begin{array}{rcl}
\overbrace{x_1 x_2 .. x_n} & = & \overbrace{x_1 .. x_{d1}} \quad \overbrace{x_{d1+1} .. x_n} \\
\hline
x_{1...n} & = & x_{1...d1} \quad \times \quad x_{d1+1...n} \\
\hline
 & & \overbrace{x_1 .. x_{d2}} \quad \overbrace{x_{d2+1} .. x_{d1}} \\
x_{1...d1} & = & x_{1...d2} \quad \times \quad x_{d2+1...d1} \\
\hline
 & & \overbrace{x_{d1+1} .. x_{d3}} \quad \overbrace{x_{d3+1} .. x_n} \\
x_{d1+1...n} & = & x_{d1+1...d3} \quad \times \quad x_{d3+1...n} \\
\hline
 & \cdots &
\end{array}
$$

where $x_{i...j} = [x_i x_{i+1} .. x_j]_L$.

For instance, consider the term $x_1 x_2 x_3 x_4 x_5$. The proposed quadrification process would operate in the following way:

$$
\begin{array}{rcl}
\overbrace{x_1 x_2 x_3 x_4 x_5} & & \overbrace{x_1 x_2 x_3} \quad \overbrace{x_4 x_5} \\
\hline
y_1 & = & y_2 \quad \times \quad y_3 \\
\hline
 & & \overbrace{x_1 x_2} \quad \overbrace{x_3} \\
y_2 & = & y_4 \quad \times \quad x_3 \\
\hline
 & & \overbrace{x_4} \quad \overbrace{x_5} \\
y_3 & = & x_4 \quad \times \quad x_5 \\
\hline
 & & \overbrace{x_1} \quad \overbrace{x_2} \\
y_4 & = & x_1 \quad \times \quad x_2 \\
\hline
\end{array}
$$

So, this quadrification is performed by recursively decomposing each product $x_i .. x_j$ into two products $x_i .. x_d$ and $x_{d+1} .. x_j$. Of course, there are many ways to choose the position of d. Sahnidis et al. [20, 22] use what they call rAI, "recursive interval arithmetic", which is a recursive quadrification where $d = j - 1$. We use the middle heuristic Qmid, where $d = (i + j)/2$, to obtain balanced degrees on the generated terms. Note that $[E]_{RI}$ contains the set of equalities that transforms a product term E into a set of quadratic identities.

The second step consists in a *Bilinear relaxation* $[[\mathcal{C}]_{RI}]_R$ of all the quadratic identities in $[\mathcal{C}]_{RI}$ with the bilinear relaxations introduced in sub-section 3.1.

Sherali and Tuncbilek [27] have proposed a direct reformulation/linearization technique (RLT) of the whole polynomial constraints without quadrifying the constraints. Applying RLT on the product term $x_1 x_2 .. x_n$ generates the following n-ary inequalities [2] :

$$\prod_{i \in J_1} (x_i - \underline{x_i}) \prod_{i \in J_2} (\overline{x_i} - x_i) \geq 0, \forall J_1, J_2 \subseteq \{1, \ldots, n\} : |J_1 \cup J_2| = n \qquad (6)$$

where $\{1, \ldots, n\}$ is to be understood as a multi-set and where J_1 and J_2 are multi-sets.

Proposition 1 bounds the number of new variables and relaxations respectively generated by the quadrification and RLT process on the product term (5).

Proposition 1.
Let $T \equiv x_1 x_2 \ldots x_n$ be some product of degree $n \geq 1$ with n distinct variables. The RLT of T will generate up to $(2^n - n - 1)$ new variables and 2^n inequalities whereas the quadrification of T will only generate $(n - 1)$ new variables and $4(n - 1)$ inequalities.

Proof: The number of terms of length i is clearly the number of combinations of i elements within n elements, that's to say C_n^i. In the RLT relaxations (6), we generate new variables for all these combinations. Thus, the number of variables is bounded by $\sum_{i=2\ldots n} C_n^i = \sum_{i=0\ldots n} C_n^i - n - 1$, that's to say $2^n - n - 1$ since $\sum_{i=0\ldots n} C_n^i = 2^n$. In (6), Dietmaier considers for each variable alternatively lower and upper bound, thus there are 2^n new inequalities.

For the quadrification process, the proof can be done by induction. For $n = 1$, the formula is true. Now, suppose that for length i (with $1 \leq i < n$), $(i - 1)$ new variables are generated. For $i = n$, we can split the term at the position d with $1 \leq d < n$. It results from the induction hypothesis that we have $d - 1$ new variables for the first part, and $n - d - 1$ new variables for the second part, plus one more new variable for the whole term. So, $n - 1$ new variables are generated. Bilinear terms require four relaxations, thus we get $4(n - 1)$ new inequalities. $\qquad \square$

Sherali and Tuncbilek [26] have proven that RLT yields a tighter linearization than quadrification on general polynomial problems. However, since the number of generated linearizations with RLT grows in an exponential way, this approach may become very expensive in time and space for non trivial polynomial constraint systems.

Proposition 2 states that quadrification with bilinear relaxations provides convex and concave envelopes with any d. This property results from the proof given in [20] for the rAI heuristic.

---

[2] Linearizations proposed in RLT on the whole polynomial problem are built on every non-ordered combination of δ variables, where δ is the highest polynomial degree of the constraint system.

Proposition 2.
Let $x_1 x_2 \ldots x_n$ *be some product of degree* $n \geq 2$ *with* n *distinct positive variables* $x_i \in \mathbf{R}_+, i = 1..n$. *Then* $[[x_1 x_2 ..x_n]_{RI}]_R$ *provides convex and concave envelopes of the product term* $x_1 x_2 ..x_n$.

Generalisation for sums of products –the so-called multi-linear terms – have been studied recently [4, 23, 17, 20]. It is well known that finding the convex or concave envelope of a multi-linear term is a NP hard problem [4]. The most common method of linear relaxation of multi-linear terms is based on the simple product term. However, it is also well known that this approach leads to a poor approximation of the linear bounding of the multi-linear terms. Sherali [23] has introduced formulae for computing convex envelopes of the multi-linear terms. It is based on an enumeration of vertices of a pre-specified polyhedra which is of exponential nature. Rikun [17] has given necessary and sufficient conditions for the polyhedrality of convex envelopes. He has also provided formulae of some faces of the convex envelope of a multi-linear function. To summarize, it is difficult to characterize convex and concave envelopes for general multi-linear terms. Conversely, the approximation of "product of variables" is an effective approach; moreover, it is easy to implement [22, 21].

3.4   Univariate Terms

Here, we provide some relaxations to handle some univariate terms. An overestimation of a convex univariate function f is given by the following envelope:

$$[f(x)]_R \; = \; [f(\underline{x}) + \frac{f(\overline{x}) - f(\underline{x})}{\overline{x} - \underline{x}}(x - \underline{x}) \geq f(x)]_L \tag{7}$$

To underestimate a convex function, we could use the **sandwich** algorithm recently analyzed by Rote [18] and which has been extended by Sahinidis and Twarmalani [22, 21]. Outer estimation of concave functions is based on the following observation : if f is a concave function, then $-f$ is a convex function.

To relax general non-convex functions, splitting is required to identify the convex and concave regions where the above relaxation can be used. To avoid branching, different techniques have been proposed. In the RLT framework [24, 28] many polynomial relaxations have been proposed for bounding univariate terms. These polynomial relaxations are then linearized with RLT techniques.

4   Experimental Results

This section reports experimental results on twenty standard benchmarks on which the extended version of Quad has been evaluated. Benchmarks eco6, katsura5, katsura6, katsura7, tangents2, ipp, assur44, cyclic5, tangents0,

Table 1. Experimental results: comparing Quad and Constraint solvers

Name	n	δ	BP(Box+Quad(Qmid))			BP(Box)			Realpaver	
			$nSols$	$nSplits$	$T(s)$	$nSols$	$nSplits$	$T(s)$	$nSols$	$T(s)$
cyclic5	5	5	10(10)	650	69.61	10(10)	13373	26.33	10	291.64
eco6	6	3	4(4)	1069	15.69	4(4)	1736	3.73	4	1.26
tangents2	6	2	24(24)	197	39.06	24(24)	14104	27.92	24	16.48
assur44	8	3	10(10)	74	68.11	10(10)	15848	72.55	10	72.56
geneig	6	3	10(10)	5053	417.86	10(10)	290711	868.64	10	475.65
ipp	8	2	10(10)	34	6.82	10(10)	4649	13.96	10	16.80
katsura5	6	2	15(11)	56	10.74	41(11)	8181	12.66	12	6.69
katsura6	7	2	44(28)	503	142.85	182(24)	136597	281.43	32	191.76
kin2	8	2	10(10)	40	7.40	10(10)	3463	19.27	10	2.61
noon5	5	3	11(11)	107	19.65	11(11)	50165	58.69	11	39.01
camera1s	6	2	16(16)	8318	452.97	2(2)	3027924	–	0	–
didrit	9	2	4(4)	90	17.39	4(4)	51284	132.94	4	94.60
kinema	9	2	8(8)	221	25.36	15(7)	244040	572.42	8	268.40
katsura7	8	2	49(43)	1729	831.96	180(35)	1421408	–	44	4675.59
lee	9	2	4(4)	491	54.56	0(0)	2091946	–	0	–
reimer5	5	6	24(24)	132	79.53	24(24)	2230187	2982.92	24	734.10
stewgou40	9	4	40(40)	1538	874.64	6(6)	779925	–	4	–
yama195	60	3	3(3)	6	114.84	0(0)	4997	–	0	–
yama196	30	1	16(0)	108	31.44	0(0)	206900	–	0	–

chemequ, noon5, geneig, kinema, reimer5, camera1s were taken from Verschelde's web site [30], kin2 from [29], didrit from [5] (page 125), lee from [12], and finally yama194, yama195, yama196 from [31]. The most challenging benchmark is stewgou40 [6]. It describes a Gough-Stewart platform with variations on the initial position of the robot as well as on its geometry. The constraint system consists of 9 equations with 9 variables. They express the length of the rods as well as the distances between the connection points.

The experimental results are reported in Tables 1 and 2. Column n (resp.) shows the number of variables (resp. the maximum polynomial degree). Experimentations with BP(X), which stands for a *Branch and Prune* solver based on the X filtering algorithm, have been performed with the implementation of iCOs [3]. Quad(H) denotes the Quad algorithm where bilinear terms are relaxed with formulas (1), power terms with formulas (4) and product terms with the quadrification method; H stands for the heuristic used for decomposing terms in the quadrification process.

The relaxations of univariate functions that have been introduced in section 3.4 have not been exploited, except for the one of the power terms through (4).

The performances of the following five solvers have been investigated:

---

[3] See http://www-sop.inria.fr/coprin/ylebbah/icos

Table 2. Experimental results: comparing solvers based on different relaxations

Name	n	δ	BP(Box+Simplex)			BP(Box+Quad(Qmid))			BP(Box+Quad(rAI))		
			$nSols$	$nSplits$	$T(s)$	$nSols$	$nSplits$	$T(s)$	$nSols$	$nSplits$	$T(s)$
cyclic5	5	5	10(10)	15830	99.98	10(10)	650	69.61	10(10)	660	96.78
eco6	6	3	4(4)	1073	6.44	4(4)	1069	15.69	4(4)	1069	15.74
tangents2	6	2	24(24)	13833	170.58	24(24)	197	39.06	24(24)	197	38.75
assur44	8	3	10(10)	15550	669.83	10(10)	74	68.11	10(10)	74	68.00
geneig	6	3	10(10)	258385	3862.20	10(10)	5053	417.86	10(10)	5053	420.04
ipp	8	2	10(10)	3151	71.24	10(10)	34	6.82	10(10)	34	6.86
katsura5	6	2	41(11)	7731	87.17	15(11)	56	10.74	15(11)	56	10.70
katsura6	7	2	182(24)	134468	2071.93	44(28)	503	142.85	44(28)	503	142.47
kin2	8	2	10(10)	2849	75.20	10(10)	40	7.40	10(10)	40	7.42
noon5	5	3	11(11)	49606	427.28	11(11)	107	19.65	11(11)	107	19.51
camera1s	6	2	2(2)	607875	−	16(16)	8318	452.97	16(16)	8318	451.43
didrit	9	2	4(4)	5361	149.03	4(4)	90	17.39	4(4)	90	17.38
kinema	9	2	14(6)	93248	1885.50	8(8)	221	25.36	8(8)	221	24.98
katsura7	8	2	37(3)	353735	−	49(43)	1729	831.96	49(43)	1729	830.86
lee	9	2	4(4)	129374	3695.48	4(4)	491	54.56	4(4)	491	54.45
reimer5	5	6	2(2)	959267	−	24(24)	132	79.53	24(24)	132	79.79
stewgou40	9	4	6(6)	115596	−	40(40)	1538	874.64	40(40)	1553	990.00
yama195	60	3	3(3)	12	41.69	3(3)	6	114.84	3(3)	6	113.92
yama196	30	1	16(0)	108	31.40	16(0)	108	31.44	16(0)	108	31.45

1. **RealPaver** : a free[4] *Branch and Prune* solver that dynamically combines optimised implementations of Box-consistency filtering and 2B-consistency filtering algorithms [8]

2. **BP(Box)**: a *Branch and Prune* solver based on Box-consistency, the ILOG[5] commercial implementation of Box-consistency

3. **BP(Box+simplex)**: a *Branch and Prune* solver based on Box-consistency and a simple linearization of the whole system without introducing outer-estimations of the nonlinear terms

4. **BP(Box+Quad(Qmid))**: a *Branch and Prune* solver which combines Box-consistency algorithm and the **Quad** algorithm where product terms are relaxed with the **Qmid** heuristic

5. **BP(Box+Quad(rAI))**: a *Branch and Prune* solver which combines Box-consistency algorithm and the **Quad** algorithm where product terms are relaxed with the **rAI** heuristic

Note that the **BP(Box+simplex)** solver implements a strategy that is close to Yamamura's approach [31].

---

[4] See http://www.sciences.univ-nantes.fr/info/perso/permanents/granvil/realpaver/-main.html

[5] See http://www.ilog.com/products/jsolver

All the solvers have been parameterised to get solutions or boxes with a precision of 10^{-8}. That's to say, the width of the computed intervals is smaller than 10^{-8}. A solution is said to be *safe* if we can prove its existence and uniqueness within the considered box. This proof is based on the well known Brouwer fix-point theorem (see [9]) and just requires a single test.

Columns nSol, nSplit and T(s) are respectively the number of found solutions, the number of branchings (or splittings) and the execution time in seconds. A "-" in the column T(s) means that the solver was unable to find all the solutions within two hours. All the computations have been performed on a PC with Pentium IV processor at 2.66Ghz. The number of solutions is followed by the number of *safe* solutions between brackets.

Table 1 displays the performances of RealPaver, BP(Box+Quad(Qmid)) and BP(Box). The benchmarks have been grouped into three sets. The first group contains problems where the Quad solver does not behave very well. These problems are quite easy to solve with Box-consistency algorithm and the overhead of the relaxation and the call to a linear solver does not pay off. The second group contains a set of benchmarks for which the Quad solver compares well with the two other constraint solvers : the Quad solver requires always much less splitting and often less time than the other solvers. In the third group, which contains difficult problems, the Quad solver outperforms the two other constraint solvers. The latter were unable to solve most of these problems within two hours whereas the Quad solver managed to find all the solutions for all but two of them in less than 8 minutes.

For instance, BP(Box) requires about 74 hours to find the four solutions of the Lee benchmark whereas Quad managed to do the job in a couple of minutes. Likewise, the Quad solver managed to find forty safe solutions of the stewgou40 benchmark in about 15 minutes whereas BP(Box) required about 400 hours.

The essential observation is that Quad solvers spend more time in the filtering step but they perform much less splitting than classical solvers. This strategy pays off for difficult problems.

Table 2 displays the performances of solvers combining Box-consistency and three different relaxation techniques. There is no significant difference between the solver based on the Qmid heuristics and the solver based on the rAI heuristics. Indeed, both heuristics provide convex and concave envelopes of the product terms.

The Quad solvers outperform Yamamura's approach for all benchmarks but yama195, which is a quasi-linear problem.

All the problems, except cyclic5 and reimer5, contain many quadratic terms and some product and power terms. cyclic5 is a pure multi-linear problem that contains only sums of products of variables. The Quad algorithm has not been very efficient for handling this problem. Of course, one could not expect an outstanding performance on this bench since product term relaxation is a poor approximation of multi-linear terms.

reimer5 is a pure power problem of degree 6, that has been well solved by the Quad algorithm. Note that Verschelde's homotopy continuation machine

[30] required about 10 minutes to solve this problem on Sparc Server 1000 and about 10 hours (on a PC equipped with a PII processor at 166Mhz) to solve `stewgou40`, another challenging problem. As opposed to the homotopy continuation method, the `Quad` solver is very simple to implement and to use. The performances on these difficult problems illustrate well the capabilities of the power relaxations.

5   Discussion

The approach introduced in this paper is related to some work done in the interval analysis community as well as to some work achieved in the optimisation community.

In the interval analysis community, Yamamura et. al. [31] have used a simple linear relaxation procedure where nonlinear terms are replaced by new variables to prove that some box does not contain solutions. No convex/concave outer-estimations are proposed to obtain a better approximation of the nonlinear terms. As pointed out by Yamamura, this approach is well adapted to quasi-linear problems : *"This test is much more powerful than the conventional test if the system of nonlinear equations consists of many linear terms and a relatively small number of nonlinear terms"* [31].

The global optimisation community worked also on solving nonlinear equation problems by transformation into an optimisation problem (see for example chapter 23 in [7]). The optimisation approach has the capability to take into account specific semantic of nonlinear terms by generating a tight outer-estimation of these terms. The pure optimisation methods are not rigorous since they do not take into account rounding errors and do not prove the existence and uniqueness of the solutions.

In this paper, we have exploited an RLT schema to take into account specific semantic of nonlinear terms. This relaxation process is incorporated in the *Branch and Prune* process [29] that exploits interval analysis and constraint satisfaction techniques to find all solutions in a given box. Experimental results show that this approach outperforms the classical constraint solvers.

A safe rounding process is a key issue for the `Quad` framework. Let's recall that the simplex algorithm is used to narrow the domain of each variable with respect to the subset of the linear set of constraints generated by the relaxation process. The point is that most implementations of the simplex algorithm are unsafe. Moreover, the coefficients of the generated linear constraints are computed with floating point numbers. So, two problems may occur in the `Quad`-filtering process:

1. The whole linearization may become incorrect due to rounding errors when computing the coefficients of the generated linear constraints ;
2. Some solutions may be lost when computing the bounds of the domains of the variables with the simplex algorithm.

We have proposed in [15] a safe procedure for computing the coefficients of the generated linear constraints. The second problem has been addressed by Neumaier [16]. He proposes a simple and cheap procedure to get a rigorous lower bound of the objective function. The incorporation of these procedures in the Quad framework will allow us to a safe use of the linear relaxations.

References

1. F.A. Al-Khayyal and J.E. Falk. Jointly constrained biconvex programming. *Mathematics of Operations Research*, pages 8:2:273–286, 1983.
2. C. Audet, P. Hansen, B. Jaumard, and G. Savard. Branch and cut algorithm for nonconvex quadratically constrained quadratic programming. *Mathematical Programming*, pages 87(1), 131–152, 2000.
3. F. Benhamou, D. McAllester, and P. Van-Hentenryck. CLP(intervals) revisited. In *Proceedings of the International Symposium on Logic Programming*, pages 124–138, 1994.
4. Y. Crama. Recognition problems for polynomial in 0-1 variables. *Mathematical Programming*, pages 44:139–155, 1989.
5. O. Didrit. *Analyse par intervalles pour l'automatique : résolution globale et garantie de problèmes non linéaires en robotique et en commande robuste*. PhD thesis, Université Parix XI Orsay, 1997.
6. Peter Dietmaier. The stewart-gough platform of general geometry can have 40 real postures. In *Advances in Robot Kinematics: Analysis and Control*, pages 1–10, 1998.
7. C.A. Floudas, editor. *Deterministic global optimization: theory, algorithms and applications*. Kluwer Academic Publishers, 2000.
8. Benhamou Frdric, Goualard Frdric, Granvilliers Laurent, and Puget Jean-Franois. Revising hull and box consistency. In *Proceedings of ICLP'99, The MIT Press*, pages 230–244, 1999.
9. Eldon R. Hansen. *Global Optimization Using Interval Analysis*. Marcel Dekker, New York, 1992.
10. H.Collavizza, F.Delobel, and M. Rueher. Comparing partial consistencies. *Reliable Computing*, pages Vol.5(3),213–228, 1999.
11. Yahia Lebbah, Michel Rueher, and Claude Michel. A global filtering algorithm for handling systems of quadratic equations and inequations. *Lecture Notes in Computer Science*, 2470:109–123, 2002.
12. T-Y Lee and J-K Shim. Elimination-based solution method for the forward kinematics of the general stewart-gough platform. In *In F.C. Park C.C. Iurascu, editor, Computational Kinematics, pages 259-267. 20-22 Mai*, 2001.
13. O. Lhomme. Consistency techniques for numeric csps. In *Proceedings of IJCAI'93*, pages 232–238, 1993.
14. A. Mackworth. Consistency in networks of relations. *Journal of Artificial Intelligence*, pages 8(1):99–118, 1977.
15. Claude Michel, Yahia Lebbah, and Michel Rueher. Safe embedding of the simplex algorithm in a csp framework. In *Proc. of 5th Int. Workshop on Integration of AI and OR techniques in Constraint Programming for Combinatorial Optimisation Problems CPAIOR 2003, CRT, Universit de Montral*, pages 210–220, 2003.
16. Arnold Neumaier and Oleg Shcherbina. Safe bounds in linear and mixed-integer programming. *Mathematical Programming, Ser. A*, pages 99:283–296, 2004.

17. A. Rikun. A convex envelope formula for multilinear functions. *Journal of Global Optimization*, pages 10:425–437, 1997.
18. G. Rote. The convergence rate of the sandwich algorithm for approximating convex functions. *Comput.*, pages 48:337–361, 1992.
19. H.S. Ryoo and V. Sahinidis. A branch-and-reduce approach to global optimization. *Journal of Global Optimization*, pages 8(2):107–139, 1996.
20. H.S. Ryoo and V. Sahinidis. Analysis of bounds for multilinear functions. *Journal of Global Optimization*, pages 19:403–424, 2001.
21. V. Sahinidis and M. Twarmalani. Baron 5.0 : Global optimisation of mixed-integer nonlinear programs. Technical report, University of Illinois at Urbana-Champaign, Department of Chemical and Biomolecular Engeneering, 2002.
22. V. Sahinidis and M. Twarmalani. Global optimization of mixed-integer programs : A theoretical and computational study. *Mathematical Programming, Ser. A*, pages 99:563–591, 2004.
23. H.D. Sherali. Convex envelopes of multilinear functions over a unit hypercube and over special discrete sets. *Acta mathematica vietnamica*, pages 22(1):245–270, 1997.
24. H.D. Sherali. Global optimization of nonconvex polynomial programming problems having rational exponents. *Journal of Global Optimization*, pages 12:267–283, 1998.
25. H.D. Sherali and C.H. Tuncbilek. A global optimization algorithm for polynomial using a reformulation-linearization technique. *Journal of Global Optimization*, pages 7, 1–31, 1992.
26. H.D. Sherali and C.H. Tuncbilek. A comparison of two reformulation-linearization technique based on linear programming relaxations for polynomial porgramming problems. *Journal of Global Optimization*, pages 10:381–390, 1997.
27. H.D. Sherali and C.H. Tuncbilek. New reformulation linearization/convexification relaxations for univariate and multivariate polynomial programming problems. *Operations Research Letters*, pages 21:1–9, 1997.
28. H.D. Sherali and H. Wang. Global optimization of nonconvex factorable programming problems. *Math. Program.*, pages 89:459–478, 2001.
29. P. Van-Hentenryck, D. Mc Allester, and D. Kapur. Solving polynomial systems using branch and prune approach. *SIAM Journal on Numerical Analysis*, pages 34(2):797–827, 1997.
30. J. Verschelde. The database of polynomial systems. Technical report, http://www.math.uic.edu/ jan/Demo/, 2003.
31. Kawata H. Yamamura K. and Tokue A. Interval solution of nonlinear equations using linear programming. *BIT*, pages 38(1):186–199, 1998.

Inter-block Backtracking: Exploiting the Structure in Continuous CSPs

Bertrand Neveu[1], Christophe Jermann[2], and Gilles Trombettoni[1]

[1] COPRIN Project, CERMICS-I3S-INRIA, 2004 route des lucioles,
06902 Sophia.Antipolis cedex, B.P. 93, France
{neveu, trombe}@sophia.inria.fr
[2] Laboratoire IRIN, Université de Nantes,
2, rue de la Houssinière, B.P. 92208,
44322 Nantes cedex 3, France
Christophe.Jermann@irin.univ-nantes.fr

Abstract. This paper details a technique, called inter-block backtracking (IBB), which improves interval solving of decomposed systems with non-linear equations over the reals.

This technique, introduced in 1998 by Bliek et al., handles a system of equations previously decomposed into a set of (small) $k \times k$ sub-systems, called blocks. All solutions are obtained by combining the solutions computed in the different blocks. The approach seems particularly suitable for improving interval solving techniques.

In this paper, we analyze into detail the different variants of IBB which differ in their backtracking and filtering strategies. We also introduce IBB-GBJ, a new variant based on Dechter's graph-based backjumping.

An extensive comparison on a sample of eight CSPs allows us to better understand the behavior of IBB. It shows that the variants IBB-BT+ and IBB-GBJ are good compromises between simplicity and performance. Moreover, it clearly shows that limiting the scope of the filtering to the blocks is very useful. For all the tested instances, IBB gains several orders of magnitude as compared to a global solving.

Keywords: intervals, decomposition, backtracking, solving sparse systems.

1 Introduction

Only a few techniques can be used to compute all the solutions to a system of continuous non-linear constraints. Symbolic techniques, such as the Groebner bases [4] and Ritt-Wu methods [19] are often very time-consuming and are limited to algebraic constraints. The continuation method, also known as the homotopy technique [12, 7], may give very satisfactory results. However, finding a solution over the reals (and not the complex numbers) is not straightforward. Moreover, using it within a constraint solving tool is difficult. Indeed, the continuation method must start from an initial system "close" to the one to be solved. This renders the automatization difficult, especially for non algebraic systems.

C. Jermann et al. (Eds.): COCOS 2003, LNCS 3478, pp. 15–30, 2005.
© Springer-Verlag Berlin Heidelberg 2005

Interval techniques are promising alternatives. They obtain good results in several fields, including robust control [10] and robotics [16]. However, it is acknowledged that systems with hundreds (sometimes tens) non-linear constraints cannot be tackled in practice.

In several applications made of non-linear constraints, systems are sufficiently sparse to be decomposed by equational or geometric techniques. CAD, scene reconstruction with geometric constraints [18, 17], molecular biology and robotics represent such promising application fields. Different techniques can be used to decompose such systems into $k \times k$ *blocks*. Equational decompositions work on the graph made of variables and equations [2, 1]. When equations model geometric constraints (e.g., distances, angles, incidences), geometric decompositions based on rigidity properties generally produce smaller blocks [11, 9].

An original approach, introduced in 1998 [2], and called in the present paper *Inter-Block Backtracking* (IBB), can be used after this decomposition phase. Following the partial order between blocks given by the decomposition, a solving process can be applied within the blocks, tackling thus systems of reduced size. IBB combines the partial solutions to construct the solutions of the problem.

Although IBB could be used with other types of solvers, we have integrated interval techniques which are general-purpose and more and more efficient. The first paper [2] presented first versions of IBB which included several backtracking schemas, along with an equational decomposition technique. Since then, several variants of IBB have been developed which had never been detailed before ([11] focussed on the geometric decomposition techniques based on flow machinery.)

Contributions

This paper details the solving phases performed by IBB with interval techniques. It brings several contributions:

- Numerous experiments have been performed on existing and new benchmarks of bigger size (between 30 and 178 equations). This leads to a more fair comparison between variants. Also, this confirms that IBB can gain several orders of magnitude in computing time as compared to interval techniques applied to the whole system. Finally, it allows us to better understand subtleties when integrating interval techniques into IBB.
- A new version of IBB is presented, based on the well-known GBJ by Dechter [6]. The experiments show that IBB-GBJ is a good compromise between previous versions.
- An inter-block interval filtering can be added to IBB. Its impact on performance is experimentally analyzed.

Contents

Section 2 gives some hypotheses about the problems that can be tackled. Section 3 recalls the principles behind IBB and interval solving. Section 4 details IBB-GBJ and the inter-block interval propagation strategy. Section 5 reports experiments performed on a sample of eight benchmarks. A discussion is given in

Section 6 on how to correct a heuristics, used inside IBB, that might lead to a loss of solutions.

2   Assumptions

IBB works on a decomposed system of equations over the reals. Any type of equation can be tackled a priori, algebraic or not. Our benchmarks contain linear and quadratic equations. IBB is used for finding **all** the solutions of a constraint system. It could be modified for global optimization (selecting the solution minimizing a given criterion) by replacing the inter-block backtracking by a classical branch and bound. Nothing has been done in this direction so far.

We assume that the systems have a finite set of solutions, that is, the variety of the solutions is 0-dimensional. This condition also holds on every sub-system (block), which allows IBB to combine together a finite set of partial solutions.

Because the conditions above are also respected for our benchmarks and because one equation can generally fix one of its variables, the system is *square*, that is, it contains as many equations as variables to be assigned; the blocks are square as well.

No more hypotheses must hold on the decomposition technique. However, since we use a structural decomposition, the system must include no redundant constraint, that is, no dependent equations. Inequalities or additional equations must be added during the solving phase in the block corresponding to their variables (as "soft" constraints in Numerica [8]), but this integration is out of the scope of this article.

For handling redundant equations, decompositions based on symbolic techniques can be envisaged [5]. These algorithms take into account the *coefficients* of the constraints, and not only the structural dependencies between variables and constraints.

Remark

In practice, the problems which can be decomposed are under-constrained and have more variables than equations. However, in existing applications, the problem is made square by assigning an initial value to a subset of variables called *input parameters*. The values of input parameters may be given by the user, read on a sketch, given by a preliminary process (e.g., in scene reconstruction [18]), or may come from the modeling (e.g., in robotics, the degrees of freedom are chosen during the design of the robot and serve to pilot it).

3   Background

First, this section briefly presents interval solving. The simplest version of IBB is then introduced on an example.

3.1   Interval Techniques

Continuous CSP

A continuous CSP $P = (V, C, I)$ contains a set of constraints C and a set of n variables V. Every variable $v_i \in V$ can take a real value in the interval $d_i \in I$; the bounds of d_i are floating-point numbers. Solving P consists in assigning variables in V to values such that all the constraints in C are satisfied.

A n-set of intervals can be represented by an n-dimensional parallelepiped called **box**. Reals cannot be represented in computer architectures, so that a solving process reduces the initial box and stops when a very small box has been obtained. Such a box is called an **atomic box** in this paper. In theory, an interval could have a width of one float at the end. In practice, the process is interrupted when all the intervals contain w_1 floats[1]. It is important to highlight that an atomic box does not necessarily contain a solution. Indeed, the process is semi-deterministic: evaluating an equation with interval arithmetic can prove that the equation has no solution (when the left and right boxes do not intersect), but cannot assert that there exists a solution in the intersection of left and right boxes.

The Interval Solver Used in IBB

We use `IlogSolver` version 5.0 and its `IlcInterval` library. `IlcInterval` implements most of the features of the language `Numerica` [8]. These libraries use several principles developed in interval analysis and in constraint programming. The interval solving process used with IBB can be summarized as follows:

1. *Bisection:* One variable is chosen and its domain is split into two intervals (the box is split along one of its dimensions). This yields two smaller sub-CSPs which are handled in sequence. This makes the solving process combinatorial.
2. *Filtering/propagation:* Local information (on constraints handled individually) or a more global one (3B) is used to reduce the current box. If the current box becomes empty, the corresponding branch (with no solution) in the search tree is cut [14, 8].
3. *Unicity test:* It is performed on the whole system of equations. It takes into account the current box B and the first and/or second derivatives of equations. When it succeeds, it finds a box B' that contains a unique solution. Also, a specific local numeric algorithm, starting from the center of B', can converge to the solution. Thus, this test generally avoids further bisection steps on B.

The three steps are iteratively performed. The process stops when an atomic box of size less than w_1 is obtained, or when the unicity test is verified on the current box.

---

[1] w_1 is a user-defined parameter. In most implementations, w_1 is a width and not a number of floats.

Propagation is performed by an AC3-like fix-point algorithm. Four types of filtering reduce the bounds of intervals (no hole is created in the current box). The `box-consistency` [8] comes from `IlcInterval`; the `2B-consistency` works in `IlogSolver`. Although algorithmically different, they both consider one constraint at a time for reducing the bounds of the implied variables (like AC3), and can be used together. The `3B-consistency` [14] uses the `2B-consistency` as sub-routine and a refutation principle (shaving) to reduce the bounds of every variable iteratively. The `bound-consistency` follows the same principle, but uses the `box-consistency` as sub-routine. A parameter w_2 is specified for the `bound` or the `3B`: a bound of a variable is not updated if the reduction is less than w_2. The w_1 parameter is also used to avoid a huge number of propagations in case of slow convergence of 2B or Box: a reduction is performed when the portion to be removed is greater than w_1.

The unicity test is implemented in `IlcInterval`. Unfortunatly, due to the implementation, it can be performed only with `Box` or `Bound`, and also cannot be called with 2B or 3B alone. This sometimes prevents us from finely analyzing the behavior of the solving.

3.2   IBB-BT

IBB works on a **Directed Acyclic Graph** of blocks (in short **DAG**) produced by any decomposition technique. A **block** i is a sub-system containing equations and variables. Some variables in i, called **input variables**, will be replaced by values during the solving of the block. The other variables are called **output variables**. A **square** block has as many equations as output variables. There exists an arc from a block i to a block j iff an equation in j involves at least one variable solved in i. The block i is called parent of j. The DAG implies a partial order in the solving performed by IBB.

Example

To illustrate the principle of IBB, we will take the 2D mechanical configuration example introduced in [2] (see Fig. 1). Various points (white circles) are con-

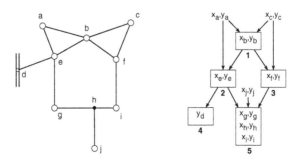

Fig. 1. Didactic problem and its DAG

nected with rigid rods (lines). Rods impose a distance constraint between two points. Point h (black circle) differs from the others in that it is attached to the rod $\langle g, i \rangle$. Finally, point d is constrained to slide on the specified line. The problem is to find a feasible configuration of the points so that all constraints are satisfied. An equational decomposition technique produces the DAG shown in Fig. 1-right.

Illustration of IBB

Respecting the order of the DAG, IBB follows one of the induced total orders, e.g., block 1, then 2, 3, 4, and 5. It first calls the interval-based solver on block 1 and obtains a first solution for x_b (the block has two solutions). Once we have this solution, we can substitute x_b by its value in the equations of subsequent blocks: 2 and 3. Then we process block 3, 4 and 5 in a similar fashion.

When a block has no solution, one has to backtrack. A chronological backtracking goes back to the previous block. IBB computes a different solution for that block and restarts to solve the blocks downstream. However, due to the chronological backtracking of this IBB-BT version, the partial order induced by the DAG is not taken into account. Indeed, in the example above, suppose block 5 had no solution. Chronological backtracking would go back to block 4, find a different solution for it, and solve block 5 again. Clearly, the same failure will be encountered again in block 5.

It is explained in [2] that the CBJ and Dynamic backtracking schemas cannot be used to take into account the structure given by the DAG. An intelligent backtracking, IBB-GPB, was introduced, based on the partial order backtracking [15, 2]. The main difficulty in implementing IBB-GPB is to maintain a set of nogoods. Moreover, any modification of IBB-GPB, for adding a feature or heuristics, such as the inter-block filtering, demands a great attention.

We present in this paper a simpler variant based on the graph-based back-Jumping (in short GBJ) by Dechter [6], and we compare it with IBB-GPB and IBB-BT.

Remarks

The reader should notice a significant difference between IBB and the backtracking schema used in finite CSPs. The domains of variables in a CSP are static, whereas the equation system in a block evolves and so does the corresponding set of solutions. Indeed, when a new solution has been selected in a parent, the corresponding variables are replaced by new values. Hence, the current block contains a new system of equations because the equations have different coefficients.

Due to interval techniques, one does not obtain a solution made of a set of scalars, but an atomic box. Thus, replacing variables from the parent blocks by constants amounts in introducing small constant intervals of width w_1 in the current block to be solved. However, the solver we use does not allow us to define constant intervals. Therefore we need to resort with a **midpoint heuristics** that replaces a constant interval by a (scalar) floating-point number comprised in it (in the "middle"). This heuristics has several significant implications on solving that are discussed in Section 6.1.

4   Use of the DAG Structure and Inter-block Filtering

The structure of the DAG can be taken into account in two ways:

- *top-down:* a *recompute condition* can sometimes avoid to compute again solutions in a block;
- *bottom-up:* when a block has no solution, one can backtrack (or backjump) to a parent block, and not necessarily to the previous block.

The following two subsections present these improvements. The third one details the inter-block filtering which can be added to all the backtracking schemas. This leads to several variants of IBB which are fully tested on our benchmarks.

4.1   The Recompute Condition

This condition can be tested in all the IBB variants, even in IBB-BT. Testing the recompute condition is not costly and leads to significant gains in performance.

The **recompute condition** states that it is useless to compute a solution in a block if the parent variables have not changed. In that case, IBB can reuse the solutions computed the last time the block has been handled. Let us illustrate when it can occur on the didactic example solved by IBB-BT.

Suppose that a first solution has been computed in block 3, and that all the solutions computed in block 4 have led to no solution. IBB-BT then backtracks on block 3 and the second position of point f is computed. When IBB goes down again to block 4, that block should normally be recomputed from scratch due to the modification of f. But x_f and y_f are not implied in equations of block 4, so that the two solutions of block 4 previously computed can be reused at this step. It is easy to avoid this useless computation by using the DAG: when IBB goes down to block 4, one checks that the parent variables x_e and y_e have not changed, so that the stored solutions can be reused.

4.2   IBB-GBJ

Six arrays are used in IBB-GBJ:

- solutions[i, j] yields the jth solution of block i.
- #sols[i] yields the number of solutions in block i.
- sol_index[i] yields the index of the current solution in block i.
- blocks_back[i] yields the set of blocks that may be the causes of failure of block i. The more recently visited block among them (i.e., the one with the highest number) is selected in case of backtracking.
- parents[i] yields the set of parent blocks of block i.
- assignment[v] yields the current value assigned to variable v.

– save_parents[i] yields the values of the variables in the parent blocks of i the last time i has been solved. This array is only used when the recompute condition is called.

IBB-GBJ can find all the solutions to a continuous CSP. Based on the DAG, the blocks are first ordered in a total order and numbered from 1 to #blocks. After an initialization phase, the while loop corresponds to the search for solutions, i being the current block. The process ends when $i = 0$, which means that all the solutions below have been found.

```
Algorithm IBB_GBJ (#blocks, solutions, parents, save_parents, assignment)

  for i = 1 to #blocks  do
     blocks_back[i] = parents[i]
     sol_index[i] = 0
     #sols[i] = 0
  end_for

  i = 1
  while (i >= 1) do

     if (Parents_changed? (i, parents, save_parents, assignment)) then
         update_save_parents (i, parents, save_parents, assignment)
         sol_index[i] = 0
         #sols[i] = 0
     end_if

     if (sol_index[i] >= #sols[i])     and
         not (next_solution(i, solutions, #sols)))
     then
         i = backjumping (i, blocks_back, sol_index)
     else  /* solutions [i, sol_index[i] ] are assigned to block i */
         assign_block (i, solutions, sol_index, assignment)
         sol_index[i] = sol_index[i] + 1
         if (i == #blocks)   then   /* total solution found */
            store_total_solution (solutions, sol_index, i)
            blocks_back[#blocks - 1] = {1...#blocks-1}
         else
            i = i + 1
         end_if
     end_if
  end_while
```

The function next_solution calls the solver to compute the next solution in the block i. If a solution has been found, the returned boolean is *true*, and the arrays solutions and #sols are updated. Otherwise, the function returns *false*.

The body corresponding to the first `else` contains actions to be performed when a solution of a block is selected. The procedure `assign_block` modifies the array `assignment` such that the values of the solution found are assigned to the variables of block i. When a total solution is found, `blocks_back[#blocks]` is updated with all the previous blocks to ensure completeness [6]. The recompute condition is checked by the function `Parents_changed?`[2].

When a block has no solution, a standard function `backjumping` returns a new level j where it is possible to backtrack without losing any solution. It is important to add in the causes of failure of block j (i.e., `blocks_back[j]`) those of block i. Indeed, those blocks are a possible cause of failure for the current value in block i.

```
function backjumping (i, in-out blocks_back, in-out sol_index)

if blocks-back[i] then
    j = more_recent (blocks_back[i])
    blocks_back[j] = blocks_back[j] U blocks_back[i] \ {j}
else
    j = 0
end_if

for k = j+1 to i do
    blocks_back[k] = parents[k]
    sol_index[k] = 0
end_for

return j
```

Favoring the Current Value

The main drawback of algorithms based on backjumping is that the work performed by the blocks between i and j is lost. When those blocks are handled again, one selects first the current value of a variable, instead of traversing the domain from the beginning. Since the domains are dynamic with IBB (the solutions of a block change when new input values are given to it), this improvement can be performed only when the recompute condition allows IBB to reuse the previous solutions.

This heuristics has been added to IBB-GBJ[3]. However, probably due to the remark above, the gains in performance obtained by the heuristics are small and are not detailed in the description of experiments (see Section 5).

---

[2] A simple way to discard this improvement is to force `Parents_changed?` to always return *true*.

[3] The algorithm must manage another index in addition to `sol_index`.

4.3   Inter-block Filtering

Contrary to the features above related to backjumping, *inter-block filtering* (in short *ibf*) is specific to interval techniques. *ibf* can thus be incorporated into any variant of IBB using an interval-based solver.

In finite CSP instances, it has generally been observed that, during the solving, performing filtering on all the remaining problem is fruitful. Therefore we decided to embedd an inter-block filtering in IBB: instead of limiting the filtering process (based on 2B, 3B, Box or Bound in our tool) to the current block, we have extended the scope of filtering to all the variables.

More precisely, before solving a block i, one forms a subsystem of variables and equations extracted from the following blocks:

1. take the set $B = \{i...\#blocks\}$ containing the blocks not yet "instantiated",
2. keep in B only the blocks connected to i in the DAG[4].

Then, the bisection is applied only on block i while the filtering process can be run on all the variables of blocks in B.

To illustrate *ibf*, let us consider the DAG of the didactic example. When block 1 is solved, all the blocks are considered by *ibf* since they are all connected to block 1. Thus, any interval reduction in block 1 can imply a reduction in any variable of the system. When block 2 is solved, a reduction can have an influence on blocks 3, 4, 5 for the same reasons. (Notice that block 3 is not downstream to block 2.) When block 3 is solved, a reduction can have an influence on blocks 5 only. Indeed, after having removed blocks 1 and 2, block 3 and 4 do not belong to the same connected component. In fact, no propagation can reach block 4 since the parent variables of block 5 which are in block 2 have an interval of width at most w_1 and thus cannot be still reduced.

Remark

One must pay attention to the way *ibf* is incorporated in IBB–GBJ. Indeed, the reductions induced by the previous blocks must be regarded as possible causes of failure. This modification is not detailed and we just illustrate the point on the DAG of the didactic example. If no solution is found in block 3, IBB with *ibf* must go back to block 2 and not to block 1. Indeed, when block 2 had been solved, a reduction could have propagated on block 3 (through 5).

5   Experiments

Exhaustive experiments have been performed on 8 benchmarks made of geometric constraints. They compare different variants of IBB and interval solving applied to the whole system (called *global solving* below).

---

[4] The orientation of the DAG is forgotten at this step, that is, the arcs of the DAG are transformed in non-directed edges, so that the filtering can apply on "brother" blocks.

5.1   Benchmarks

Some of them are artificial problems, mostly made of quadratic distance constraints. Mechanism and Tangent have been found in [13] and [3]. Chair is a realistic assembly made of 178 equations from a large variety of geometric constraints: distances, angles, incidence, parallelisms, orthogonalities, etc.

The domains have been selected around a given solution and lead to radically different search spaces. Note that a problem defined with large domains is generally similar to assign $]-\infty, +\infty[$ to every variable.

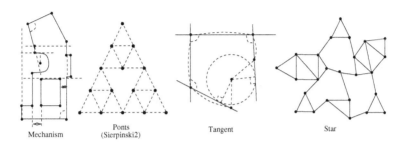

Mechanism   Ponts   Tangent   Star
                (Sierpinski2)

Fig. 2. 2D benchmarks: general view

Table 1. Details on the benchmarks: type of decomposition method. (Dec., see Section 1); number of equations (Size); Size of blocks (Size Dec.)- NxK means N blocks of size K - # of solutions with the four types of domains selected: tiny (width = 0.1), small (1), medium (10), large (100)

Dim	GCSP	Dec.	Size	Size Dec.	Ti.	Small	Med.	Large
2D	Mechanism	equ.	98	98 = 1x10, 2x4, 27x2, 26x1	1	8	48	448
	Ponts	equ.	30	30 = 1x14, 6x2, 4x1	1	15	96	128
	Sierpinski3	geo.	84	124 = 44x2, 36x1	1	8	96	138
	Tangent	geo.	28	42 = 2x4, 11x2, 12x1	4	16	32	64
	Star	equ.	46	46 = 3x6, 3x4, 8x2	1	4	8	8
3D	Chair	equ.	178	178 = 1x15,1x13,1x9,5x8,3x6,2x4,14x3,1x2,31x1	6	6	18	36
	Hourglass	geo.	29	39 = 2x4, 3x3, 2x2, 18x1	1	1	2	8
	Tetra	equ.	30	30 = 1x9, 4x3, 1x2, 7x1	1	16	68	256

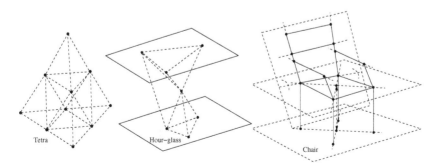

Tetra   Hour-glass   Chair

Fig. 3. 3D benchmarks: general view

Sierpinski3 is the fractal Sierpinski at level 3, that is, 3 Sierpinski2 put together. The corresponding equation system would have about 2^{40} solutions, so that the initial domains are limited to a width 0.1 (tiny), 0.8 (small), 0.9 (medium), 1 (large).

5.2   Choice of Filtering

With the aim of not handicapping the global solving, we select the best filtering algorithms by performing tests on two benchmarks of medium size. Several widths have been tried for w_1 and w_2 (see Table 2).

Table 2. Comparison of different partial consistencies. All the times are obtained in seconds on a `PentiumIII 935 Mhz` with a Linux operating system. The best results appear in bold-faced. A 0 in column w_2 means that the lines 1 and 4 report results obtained by 2B, Box, or 2B+Box. Otherwise, when w_2 is 1e-2 or 1e-4, the corresponding lines report results obtained by 3B, bound, or 3B+bound. Cells containing *sing.* (singularity) indicate that multiple solutions are obtained and lead to a combinatorial explosion (see Section 6)

	w_2	w_1	2B/3B	Box/Bound	2B+Box/3B+Bound
Ponts	0	1e-6	*sing.*	264	**29**
		1e-8	*sing.*	292	**32**
		1e-10	*sing.*	278	**32**
	1e-2	1e-6	116	2078	309
		1e-8	2712	2642	1303
		1e-10	13565	2652	5570
	1e-4	1e-6	84	>54000	523
		1e-8	4413	>54000	5274
Tangent	0	1e-6	*sing.*	547	81
		1e-8	*sing.*	553	82
		1e-10	*sing.*	562	86
	1e-2	1e-6	**26**	265	91
		1e-8	**35**	270	94
		1e-10	**60**	266	93
	1e-4	1e-6	51	2516	369
		1e-8	68	2535	393

Clearly, 2B+Box and 3B outperfom the other combinations. All the following tests have been performed with these two filtering techniques.

5.3   Main Tests

The main conclusions about the tests reported in Table 3 are the following:

- IBB always outperforms the global solving, which highlights the interest of exploiting the structure. One, two or three orders of magnitude can be gained

Table 3. Results of experiments. `BT+` is `IBB-BT` with the recompute condition. For every algorithm and every domain size, times are given either without inter-block filtering (¬IBF) or with IBF. All the times are obtained in seconds on a `PentiumIII 935 Mhz` with a Linux operating system. The reported times are obtained with `2B+Box` which is often better than 3B. The lines `3B(GBJ)` report times with `IBB-GBJ` and 3B when it is competitive

		Tiny		Small		Medium		Large	
		¬IBF	IBF	¬IBF	IBF	¬IBF	IBF	¬IBF	IBF
	Global	XXS		XXS		XXS		XXS	
Chair	BT	3.3	XXS	3.2	XXS	9.4	XXS	12.4	XXS
	BT+	2.4	XXS	2.3	XXS	4.5	XXS	4.7	XXS
	GBJ	2.4	XXS	2.3	XXS	4.5	XXS	4.7	XXS
	Global	XXS		XXS		XXS		XXS	
Mechanism	BT	0.17	14.1	0.6	15.0	2.8	18.7	13.3	32.8
	BT+	0.11	14.1	0.4	13.6	2.6	17.2	13.1	30.6
	GBJ	0.10	14.1	0.4	13.5	2.6	17.3	13.1	30.4
	GPB	0.10	14.2	0.4	13.3	2.7	17.4	13.1	30.5
	3B(GBJ)	0.23	0.68	1.7	2.3	9.7	11	83	88
	Global	0.73		32		82		110	
Ponts	BT	0.16	0.63	2.38	4.2	6.5	10.6	9.1	14.6
	BT+	0.16	0.63	2.36	4.2	6.1	10.2	8.8	14.7
	GBJ	0.17	0.58	2.35	4.1	6.0	10.4	8.7	14.4
	GPB	0.22	0.61	2.37	4.1	6.3	10.4	8.7	14.4
	3B(GBJ)	0.3	0.6	12	15	25	31	49	59
	Global	0.12		1.89		1.47		22.77	
Hour-glass	BT+	0.03	0.88	0.03	1.64	0.06	1.00	0.06	1.21
	GBJ	0.04	0.75	0.03	1.60	0.02	0.83	0.06	1.19
	GPB	0.05	0.73	0.03	1.61	0.05	0.88	0.05	1.15
	3B(GBJ)	0.03	0.3	0.05	0.6	0.05	0.2	0.1	0.4
	Global	3.1		>54000		>54000		>54000	
Sierpinski3	3B(BT)	0.1	1.32	12.3	160	96	788	136	1094
	3B(BT+)	0.1	1.32	12.7	160	67	703	93	928
	3B(GBJ)	0.1	1.32	12	166	61	682	85	916
	Global	0.5		35		39		46	
Tangent	BT+	0.05	1.26	0.11	1.89	0.13	7.63	0.20	8.15
	BJ	0.07	1.17	0.11	1.89	0.14	7.69	0.19	8.00
	GPB	0.07	1.19	0.10	1.93	0.11	7.69	0.22	8.04
	3B(GBJ)	0.2	0.7	0.2	1.3	0.2	1.3	0.3	1.7
	Global	2.15		92		197		406	
Tetra	BT+	0.14	0.74	1.08	4.00	2.37	7.01	4.73	13.56
	GBJ	0.14	0.67	1.10	3.87	2.30	6.80	4.74	13.20
	GPB	0.16	0.65	1.11	3.90	2.29	6.71	4.72	13.19
	Global	8.7		2908		2068		1987	
Star	BT	9.96	70	40.2	137	81.4	241	80.3	240
	BT+	9.96	70	29.5	99.6	78.1	102	78	102
	GBJ	9.96	70	29.1	99.6	77.9	102	77.9	102
	GPB	9.96	70	29.4	99.6	49.3	102	49	102

in performance. Even with tiny domains, the gains can be significant (see `Sierpinski3`)[5].

- The inter-block filtering is always counter-productive and sometimes very bad (see `Tangent`). Several lines with 3B have been added to show that the loss of time of inter-block filtering is reduced with 3B.

---

[5] The global solving compares advantageously with `IBB` on the `Star` benchmark with tiny domains. This is due to a greater precision required for `IBB` to make it complete (see Section 6). With the same precision, the global solving spends 75 s to find the solutions.

Table 4. Number of backjumps with and without inter-block filtering

	IB filtering	Tiny	Small	Medium	Large
Ponts	no	0	0	1	0
	yes	0	0	0	0
Mechanism	no	3	4	0	0
	yes	0	0	0	0
Star	no	0	2	6	6
	yes	0	0	0	0
Sierpinski3	no	0	12	829	2118
	yes	0	0	5	4

– The exploitation of the `DAG` structure by the recompute condition is very useful.

Remark

Entries in Table 3 containing XXS correspond to a failure in the solving process due to `IlogSolver` when a maximum size is exceeded.

To refine our conclusions, Table 4 reports statistics made on the number of backjumps performed by `IBB-GBJ`. Note that no backjumps have been observed with the other four benchmarks.

These experiments highlight a significant result. Most of the backjumps disappear with the use of inter-block filtering, which reminds similar results observed in finite CSPs. However, the price paid by inter-block filtering for removing these backjumps does not bring in good returns. `Sierpinski3`-Large highlights the trend: 2114 on 2118 backjumps are eliminated by inter-block filtering, but the algorithm is 10 times slower than `IBB-GBJ`!

6   Discussion

Two difficulties come from the use of interval techniques with `IBB`. They are detailed below.

6.1   Midpoint Heuristics

This heuristics (see Section 3.2) is not satisfactory because solutions might be lost, making the whole process incomplete[6]. When the midpoint heuristics had been introduced [2], our examples were small, and the case had not occurred. Since then, it has occurred with `Star`, `Sierpinski3` and `Chair`. The problem has been fixed with `Star` and `Sierpinski3` by increasing the precision (i.e.,

---

[6] Its correctness can be ensured by a final and quasi-immediate filtering process performed on the whole equation system, where the domains form an atomic box.

decreasing w_1). Ad-hoc modifications of the equation system must have been made to fix the problem on `Chair`.

The clean solution consists in introducing constant intervals in equations instead of the midpoints (which is not currently possible with `IlogSolver`). We think that the overcost in time would be negligible with the filtering/bisection solving schema. On the contrary, unicity tests may become inefficient because computing the inverse of a jacobian matrix including intervals (instead of scalars) can lead to large overestimates of the intervals.

6.2   Dealing with Multiple Solutions

Another limit of interval techniques is worsened with `IBB`. Multiple solutions occur when several atomic boxes are close to each other: only one contains a solution and the others are not discarded by filtering. Even when the number of multiple solutions is small, the multiplicative effect due to `IBB` (the partial solutions are combined together) may render the problem intractable.

An ad-hoc solution consists in improving the precision (i.e., reducing w_1), which fixes some cases. Mixing several filtering techniques, such as `2B+Box`, also reduces the phenomenon (The *sing.* entries in Table 2 with `2B` come from this phenomenon.) We have implemented a first way to detect multiple solutions. In this case, we select only one of them. This has solved the problem in most cases. A few pathologic cases remain due to an interaction with the midpoint heuristics. Taking the union of the multiple solutions should be more robust.

7   Conclusion

This paper has detailed the generic inter-block backtracking framework to solve decomposed continuous CSPs. We have implemented three backtracking schemas (chronological `BT`, `GBJ`, partial order backtracking). Every backtracking schema can incorporate a recompute condition that avoids sometimes a useless call to the solver. Every schema can also use an inter-block filtering.

Series of exhaustive tests have been performed on a sample of benchmarks of acceptable size and made of non-linear equations. First, all the variants of `IBB` can gain several orders of magnitude as compared to solving the constraint system globally. Second, exploiting the structure of the `DAG` with the recompute condition is very useful whereas a more sophisticated exploitation (backjumping) only improves slightly the performance of IBB. However, it might lead to important gains while never producing an overhead. This leads us to propose the `IBB-GBJ` version presented in this paper.

Another clear result of this paper is that inter-block filtering is counterproductive. This highlights that a global filtering which does not take the structure into account makes a lot of useless work.

The next step of our work is to deal with small constant intervals to discard the midpoint heuristics and make our implementation more robust.

References

1. S. Ait-Aoudia, R. Jegou, and D. Michelucci. Reduction of constraint systems. In *Compugraphic*, 1993.
2. Christian Bliek, Bertrand Neveu, and Gilles Trombettoni. Using graph decomposition for solving continuous csps. In *Principles and Practice of Constraint Programming, CP'98*, volume 1520 of *LNCS*, pages 102–116. Springer, 1998.
3. W. Bouma, I. Fudos, C.M. Hoffmann, J. Cai, and R. Paige. Geometric constraint solver. *Computer Aided Design*, 27(6):487–501, 1995.
4. B. Buchberger. An algebraic method in ideal theory. In *Multidimensional System Theory*, pages 184–232. Reidel Publishing Co., 1985.
5. B. Mourrain D. Bondyfalat and T. Papadopoulo. An application of automatic theorem proving in computer vision. In *2nd International Workshop on Automated Deduction in Geometry, Springer-Verlag*, 1999.
6. Rina Dechter. Enhancement schemes for constraint processing: Backjumping, learning, and cutset decomposition. *Artificial Intelligence*, 41(3):273–312, January 1990.
7. C. Durand. *Symbolic and Numerical Techniques For Constraint Solving*. PhD thesis, Purdue University, 1998.
8. Pascal Van Hentenryck, Laurent Michel, and Yves Deville. *Numerica : A Modeling Language for Global Optimization*. MIT Press, 1997.
9. Christoph Hoffmann, Andrew Lomonossov, and Meera Sitharam. Finding solvable subsets of constraint graphs. In *Proc. Constraint Programming CP'97*, pages 463–477, 1997.
10. L. Jaulin, M. Kieffer, O. Didrit, and E. Walter. *Applied Interval Analysis*. Springer-Verlag, 2001.
11. Christophe Jermann, Gilles Trombettoni, Bertrand Neveu, and Michel Rueher. A constraint programming approach for solving rigid geometric systems. In *Principles and Practice of Constraint Programming, CP 2000*, volume 1894 of *LNCS*, pages 233–248, 2000.
12. E. Lahaye. Une méthode de résolution d'une catégorie d'équations transcendantes. *Compte-rendu des Séances de L'Académie des Sciences*, 198:1840–1842, 1934.
13. R.S. Latham and A.E. Middleditch. Connectivity analysis: A tool for processing geometric constraints. *Computer Aided Design*, 28(11):917–928, 1996.
14. O. Lhomme. Consistency techniques for numeric csps. In *IJCAI*, pages 232–238, 1993.
15. D.A. McAllester. Partial order backtracking. Research Note, Artificial Intelligence Laboratory, MIT, 1993. ftp://ftp.ai.mit.edu/people/dam/dynamic.ps.
16. Jean-Pierre Merlet. Optimal design for the micro robot. In *in IEEE Int. Conf. on Robotics and Automation*, 2002.
17. Gilles Trombettoni and Marta Wilczkowiak. Scene Reconstruction based on Constraints: Details on the Equation System Decomposition. In *Proc. International Conference on Constraint Programming, CP'03*, volume LNCS 2833, pages 956–961, 2003.
18. Marta Wilczkowiak, Gilles Trombettoni, Christophe Jermann, Peter Sturm, and Edmond Boyer. Scene Modeling Based on Constraint System Decomposition Techniques. In *Proc. International Conference on Computer Vision, ICCV'03*, 2003.
19. W. Wu. Basic principles of mechanical theorem proving in elementary geometries. *J. Automated Reasoning*, 2:221–254, 1986.

Convex Programming Methods for Global Optimization

J.N. Hooker

GSIA, Carnegie Mellon University, Pittsburgh, USA
jh38@andrew.cmu.edu

November 2003, Revised January 2004

Abstract. We describe four approaches to solving nonconvex global optimization problems by convex nonlinear programming methods. It is assumed that the problem becomes convex when selected variables are fixed. The selected variables must be discrete, or else discretized if they are continuous. We first survey some existing methods: disjunctive programming with convex relaxations, logic-based outer approximation, and logic-based Benders decomposition. We then introduce a branch-and-bound method with convex quasi-relaxations (BBCQ) that can be effective when the discrete variables take a large number of real values. The BBCQ method generalizes work of Bollapragada, Ghattas and Hooker on structural design problems. It applies when the constraint functions are concave in the discrete variables and have a weak homogeneity property in the continuous variables.

We address global optimization problems that become convex when selected variables are fixed. If these variables are discrete, the constraints can be reformulated as logical disjunctions of convex constraints. If some of the selected variables are not discrete, we discretize them in order to obtain an approximate global solution.

The motivation for this approach is to take advantage of highly developed nonlinear programming methods for convex problems, as well as branch-and-bound methods for discrete problems. A branch-and-bound method chooses the appropriate disjunct in each constraint. Nonlinear programming is applied to the convex subproblem that results when the disjuncts are chosen.

We present four variations of this general approach.[1] Two of them are most practical when the discrete variables do not take a large number of possible values: (a) disjunctive programming with convex relaxations, and (b) logic-based outer approximation. The disjunctive programming model can also be solved as a mixed integer/nonlinear programming (MINLP) problem. When there are a large number of discrete values, as when some discrete variables represent discretized continuous variables, one can turn to methods that do not require explicit representation of the disjunctions: (c) logic-based Benders decomposi-

---

[1] A longer version of this paper, available at web.gsia.cmu.edu/jnh/cocos03.pdf, presents examples of all four methods.

C. Jermann et al. (Eds.): COCOS 2003, LNCS 3478, pp. 46–60, 2005.

tion, and (d) branch and bound with convex quasi-relaxations (BBCQ). The convergence rate of the Benders method depends heavily on the problem structure, however. BBCQ is intended for problems in which the discrete variables are real-valued. It does not rely on decomposition but requires that the constraint functions satisfy certain properties.

This paper begins with a summary of the first three methods, which are developed elsewhere. It then introduces the BBCQ method as a formalization and generalization of a technique applied by Bollapragada, Ghattas and Hooker to structural design problems [1]. This application is presented at the end of the paper as an illustration of disjunctive programming and BBCQ.

1   General Form of the Problem

We solve problems of the form

$$
\begin{aligned}
\min \quad & x_1 \\
\text{subject to} \quad & g^j(x, y_j) \leq 0, \ \ j \in J \\
& L(y) \\
& x \in \mathbb{R}^n, \ \ y_j \in Y_j, \ j \in J
\end{aligned}
\tag{1}
$$

where $g^j(x, y_j)$ is a vector of functions and $L(y)$ is a logical constraint on possible values of the discrete variables y_j. If some of the y_j are continuous, we discretize them by converting Y_j to a finite set. We assume that when each y_j is fixed to some $\bar{y}_j \in Y_j$ we obtain the convex subproblem:

$$
\begin{aligned}
\min \quad & x_1 \\
\text{subject to} \quad & g^j(x, \bar{y}_j) \leq 0, \ \ j \in J \\
& x \in \mathbb{R}^n
\end{aligned}
\tag{2}
$$

It is convex in the sense that each $g^j(x, \bar{y}_j)$ is a vector of convex functions of x.

We assume without loss of generality that the objective function is a single variable x_1, since x_1 can be defined in the constraints. We also suppose that each constraint contains only one discrete variable y_j. Many problems naturally occur in this form. Problems that do not can in principle be put into this form by a change of variables. Thus a constraint $g^j(x, y_1, \ldots, y_m) \leq 0$ can be written $g^j(x, y^j) \leq 0$, where $y^j = (y_1^j, \ldots, y_m^j)$ is regarded as a single variable. The variables y^j can now be related by the logical constraints $y^j = y^1$ for all $j \in J$. For instance, the constraints $x + y_1 + y_2 \geq b$ and $x + y_2 + y_3 \geq b$ can be rewritten $x + y_1^1 + y_2^1 \geq b$ and $x + y_2^2 + y_3^2 \geq b$ by adding the constraint $y_2^1 = y_2^2$.

2   Disjunctive Formulation

A straightforward but generally impractical way to solve (1) is by a branch-and-bound method that branches on the y_j and solves a continuous relaxation of the problem at each node of the branching tree. The difficulty is that these continuous problems are in general nonconvex.

To obtain convex relaxations, we write (1) as a *disjunctive programming problem* by creating a disjunct for each possible value of y_j.

$$\min \quad x_1$$
$$\text{subject to} \quad \bigvee_{v \in Y_j} \begin{bmatrix} y_j = v \\ g^j(x,v) \le 0 \end{bmatrix}, \quad j \in J \tag{3}$$
$$L(y)$$
$$x \in \mathbb{R}^n$$

The functions $g^j(x,v)$ are convex because the second argument is fixed. They may also simplify in form. In some cases singularities disappear, as for example when

$$g^j(x, y_j) = \begin{bmatrix} x_1 - 1/y_1 \\ x_1 - x_2 \end{bmatrix} \le \begin{bmatrix} 0 \\ 0 \end{bmatrix}$$

can be written simply $x_1 - x_2 \le 0$ for $y_j = 0$.

3   Disjunctive Programming with Convex Relaxations

A branch-and-bound method can be practical for the disjunctive programming problem (3) when it is possible to devise a convex relaxation at each node of the search tree. Two such relaxations, based on big-M and convex hull formulations, are presented here.

Branch and bound proceeds by branching on the alternatives in the disjunctions of (3). At each node of the search tree, some disjuncts have been selected by prior branching, and these are imposed as constraints. The disjunctions on which the algorithm has not yet branched are relaxed. A lower bound is obtained by solving a convex problem that minimizes x_1 subject to the imposed disjuncts and the relaxed disjunctions. The lower bound is used to prune the search as is normally done in branch-and-bound search (see [9, 11] for details).

A closely related approach is to apply an MINLP method to a 0-1 model of the disjunctive model (3), which results from imposing an integrality condition on either the big-M or the convex hull relaxation of (3).

The *big-M relaxation* introduces a variable β_{jv} for each $v \in Y_j$, where $\beta_{jv} = 1$ is interpreted as indicating $y_j = v$. It is assumed that there are bounds $x^L \le x \le x^U$ on x. Let $L(\beta)$ be an inequality encoding of the logical constraints $L(y)$ [3]. The big-M relaxation of (3) is:

$$\min \quad x_1$$
$$\text{subject to} \quad g^j(x,v) \le M^{jv}(1 - \beta_{jv}), \quad \text{all } v \in Y_j, \ j \in J$$
$$\sum_{v \in Y_j} \beta_{jv} = 1, \ \beta_{jv} \ge 0, \quad \text{all } v \in Y_j, \ j \in J \tag{4}$$
$$L(\beta), \ x^L \le x \le x^U$$
$$0 \le \beta_{jv} \le 1, \quad \text{all } v \in Y_j, j \in J$$

where M^{jv} is a vector of valid upper bounds on the component functions of $g^j(x,v)$, given that $x^L \leq x \leq x^U$. This relaxation is clearly convex.

One can solve (3) by using relaxation (4) at each node, where J in (4) corresponds to the set of disjunctions on which the algorithm has not yet branched. Alternatively, one can apply an MINLP algorithm to the 0-1 model obtained by replacing $\beta_{jv} \in [0,1]$ in (4) with $\beta_{jv} \in \{0,1\}$, where J corresponds to the original set of disjunctions.

The bounds M^{jv} should be the tightest that can be practicably obtained. One valid bound is

$$M_i^{jv} = \max_{x^L \leq x \leq x^U} \left\{ g_i^j(x,v) \right\} \tag{5}$$

but the tightest bound is

$$M_i^{jv} = \max_{v' \in Y_j \setminus \{v\}} \left\{ \max_{x^L \leq x \leq x^U} \left\{ g_i^j(x,v) \mid g^j(x,v') \leq 0 \right\} \right\}$$

A second convex relaxation for (3), based on convex hull descriptions of the disjunctions, was developed by Stubbs and Mehrotra [14] and Grossmann and Lee [7]. It is generally tighter than the big-M relaxation but requires that we introduce for each disjunction j a new continuous variable x^{jv} for each $v \in Y_j$.

The *convex hull relaxation* for a disjunction

$$\bigvee_{v \in Y_j} g^j(x,v) \leq 0 \tag{6}$$

can be derived as follows. We assume that x and g^j are bounded; that is, $x \in [x^L, x^U]$, and $g^j(x) \in [-L, L]$ for $x \in [x^L, x^U]$. We wish to characterize all points x that can be written as a convex combination of points $\hat{x}^{jv}$ that respectively satisfy the disjuncts of (6). Thus we have

$$x = \sum_{v \in Y_j} \beta_{jv} \hat{x}^{jv}$$
$$g^j(\hat{x}^j, v) \leq 0, \text{ all } v \in Y_j$$
$$x^L \leq \hat{x}^j \leq x^U$$
$$\sum_{v \in Y_j} \beta_{jv} = 1, \ \beta_{jv} \geq 0, \text{ all } v \in Y_j$$

Using the change of variable $x^{jv} = \beta_{jv}\hat{x}^{jv}$, we obtain the relaxation

$$x = \sum_{v \in Y_j} x^{jv}$$
$$g^j\left(\frac{x^{jv}}{\beta_{jv}}, v\right) \leq 0, \text{ all } v \in Y_j$$
$$\beta_{jv} x^L \leq x^{jv} \leq \beta_{jv} x^U, \text{ all } v \in Y_j \tag{7}$$
$$\sum_{v \in Y_j} \beta_{jv} = 1, \ \beta_{jv} \geq 0, \text{ all } v \in Y_j$$

The function $g^j(x^{jv}/\beta_{jv}, v)$ is in general nonconvex, but a classical result of convex analysis (e.g. [8]) implies that one can restore convexity by multiplying the second constraint of (7) by β_{jv}. A theorem very similar to the following is proved in [14] (see also [2]).

Theorem 1. *Consider the set S consisting of all (x, β) with $\beta \in [0,1]$ and $x \in [\beta x^L, \beta x^U]$. If $g(x)$ is convex and bounded for $x \in [\beta x^L, \beta x^U]$, then*

$$h(x, \beta) = \begin{cases} \beta g(x/\beta) & \text{if } \beta > 0 \\ 0 & \text{if } \beta = 0 \end{cases}$$

is convex and bounded on S.

Proof. To show convexity of $h(x, \beta)$ we arbitrarily choose $(x^1, \beta_1), (x^2, \beta_2) \in S$. Supposing first that $\beta_1, \beta_2 > 0$, we have convexity since

$$h\left(\alpha x^1 + (1-\alpha)x^2, \alpha\beta_1 + (1-\alpha)\beta_2\right)$$
$$= (\alpha\beta_1 + (1-\alpha)\beta_2)\, g\left(\frac{\alpha x^1 + (1-\alpha)x^2}{\alpha\beta_1 + (1-\alpha)\beta_2}\right)$$
$$= (\alpha\beta_1 + (1-\alpha)\beta_2)\, g\left(\frac{\alpha\beta_1}{\alpha\beta_1 + (1-\alpha)\beta_2}\frac{x^1}{\beta_1} + \frac{(1-\alpha)\beta_1}{\alpha\beta_1 + (1-\alpha)\beta_2}\frac{x^2}{\beta_2}\right)$$
$$\leq (\alpha\beta_1 + (1-\alpha)\beta_2)\left[\frac{\alpha\beta_1}{\alpha\beta_1 + (1-\alpha)\beta_2}g\left(\frac{x^1}{\beta_1}\right) + \frac{(1-\alpha)\beta_1}{\alpha\beta_1 + (1-\alpha)\beta_2}g\left(\frac{x^2}{\beta_2}\right)\right]$$
$$= \alpha h\left(x^1, \beta_1\right) + (1-\alpha)h\left(x^2, \beta_2\right)$$

for any $\alpha \in [0,1]$, where the inequality is due to the convexity of $g(x)$. If $\beta_1 = \beta_2 = 0$, then

$$h\left(\alpha x^1 + (1-\alpha)x^2, \alpha\beta_1 + (1-\alpha)\beta_2\right) = h(0,0) = \alpha h\left(x^1, \beta_1\right) + (1-\alpha)h\left(x^2, \beta_2\right)$$

since $\beta_j x^L \leq x^j \leq \beta_j x^U$ implies $x^j = 0$. If $\beta_1 = 0$ and $\beta_2 > 0$, we have

$$h\left(\alpha x^1 + (1-\alpha)x^2, \alpha\beta_1 + (1-\alpha)\beta_2\right)$$
$$= h\left((1-\alpha)x^2, (1-\alpha)\beta_2\right) = (1-\alpha)g\left(\frac{x^2}{\beta_2}\right)$$
$$= \alpha h(0,0) + (1-\alpha)h\left(x^2, \beta_2\right)$$

Finally, $h(x, \beta) = \beta g(x/\beta)$ is bounded because $\beta \in [0,1]$, $x/\beta \in [x^L, x^U]$, and $g(x)$ is bounded for $x \in [x^L, x^U]$.

We now obtain the following convex relaxation for (3):

$$\min \quad x_1$$

$$\text{subject to} \quad x = \sum_{v \in Y_j} x^{jv}, \quad \text{all } j \in J$$

$$\beta_{jv} g^j \left(\frac{x^{jv}}{\beta_{jv}}, v \right) \leq 0, \quad \text{all } v \in Y_j, \ j \in J \tag{8}$$

$$\beta_{jv} x^L \leq x^{jv} \leq \beta_{jv} x^U, \quad \text{all } v \in Y_j, \ j \in J$$

$$\sum_{v \in Y_j} \beta_{jv} = 1, \ \beta_{jv} \geq 0, \quad \text{all } v \in Y_j, \ j \in J$$

$$L(\beta), \ x, x^{jv} \in \mathbb{R}^n, \quad \text{all } v \in Y_j, \ j \in J$$

This is not a convex hull relaxation for (3) as a whole, but it provides a convex hull relaxation of each disjunction of (3).

Since β_{jv} can vanish, it is common in practice to use the constraint

$$(\beta_{jv} + \epsilon) g^j \left(\frac{x^{jv}}{\beta_{jv} + \epsilon}, v \right) \leq 0, \quad \text{all } v \in Y_j, \ j \in J$$

The introduction of ϵ preserves convexity. Grossmann and Lee [7] suggest using $\epsilon = 10^{-4}$.

4   Logic-Based Outer Approximation

One can use linear rather than convex nonlinear relaxations by modifying the outer approximation method for MILP [4] to solve disjunctive programming problems, as shown by Türkay and Grossmann [15]. The drawback is that the linear relaxations must be updated and solved repeatedly.

Logic-based outer approximation solves a master problem containing first-order approximations of the disjuncts of (3) to obtain a value $\bar{y}$ for y. It then solves the nonlinear but convex subproblem (2) to obtain a corresponding value for x. The first-order approximations are computed about the values of x obtained in previous iterations. The process continues until optimal value of the master problem approximates the largest optimal subproblem value found so far.

Let (x^k, y^k) for $k = 1, \ldots, K$ be the solutions obtained by solving the master problem and subproblem in previous iterations. The master problem in iteration $K + 1$ can be written

$$\min \quad x_1$$

$$\text{subject to} \quad \bigvee_{v \in Y_j} \left[\begin{array}{c} y_j = v \\ g^j(x^k, v) + \nabla g^j(x^k, v)(x - x^k) \leq 0, \\ \text{all } k \in \{1, \ldots, K\} \text{ with } y_j^k = v \end{array} \right], \quad \text{all } j \in J \tag{9}$$

$$L(y), \ x \in \mathbb{R}^n$$

Since the disjuncts in (9) are linear, the relaxations (4) and (8) are likewise linear. One can therefore solve (9) by applying a mixed integer programming method to a 0-1 formulation of (9). Again, either (4) or (8) can serve as a 0-1 formulation if the variables β_{jv} are treated as 0-1 variables. The solution y of (9) becomes y^{K+1}, and x^{K+1} is an optimal solution of the subproblem (2) with $\bar{y} = y^{K+1}$.

In practice it is advantageous to obtain a warm start by solving the subproblem for several values of $\bar{y}$ before solving the first master problem.

5   Logic-Based Benders Decomposition

When a constraint in the disjunctive programming formulation contains many disjuncts, the number of variables in the relaxations (4) and (8) can become quite large. This can be avoided by applying logic-based Benders decomposition to (3), which in effect uses a discrete relaxation of the problem and does not require an explicit formulation of the disjunctions [9, 12]. However, the convergence rate is unpredictable.

In logic-based Benders, the master problem consists of *Benders cuts* that contain only the discrete variables y_j. At any point in the algorithm, the Benders cuts partially describe the projection of the original problem's feasible set onto the y-space.

In iteration K the subproblem is (2) with $\bar{y}$ set to the solution y^K of the current master problem. Let λ^{Kj} be the vector of Lagrange multipliers associated with constraint j in the optimal solution of (2), and let x_1^K be the optimal value of (2). Since constraints with vanishing Lagrange multipliers are inactive in the subproblem, we can state the following: whenever $\bar{y}_j$ is set to y_j^K for all constraints j with $\lambda^{Kj} \neq 0$, the optimal value of the subproblem is still x_1^K. We generate a Benders cut that states this fact, and add it to the master problem for iteration $K + 1$:

$$
\begin{aligned}
\min \quad & z \\
\text{subject to} \quad & \bigwedge_{\substack{j \\ \lambda^{kj} \neq 0}} (y_j = y_j^k) \implies (z \geq x_1^k), \quad k = 1, \ldots, K \\
& L(y)
\end{aligned}
\tag{10}
$$

where $\implies$ means "implies." For each k the implication in (10) is the Benders cut generated in iteration k. The master problem is solved for y^{K+1}, and the process continues until the optimal value of (10) approximates the best subproblem value found so far.

The master problem can be solved by finite-domain constraint programming techniques or by converting it to an integer programming problem for an MILP solver.

In general, logic-based Benders cuts are obtained by solving the *inference dual* of the subproblem. This approach has been successfully applied to planning and scheduling problems in which the master problem is solved by integer programming and the subproblem by constraint programming [9, 10, 13]. There is little experience to date with continuous nonlinear subproblems, but decomposition is clearly more effective when most of the Lagrange multipliers vanish, since this results in stronger Benders cuts. When none of the multipliers vanish, the method reduces to exhaustive enumeration.

It is useful in practice to enhance the master problem with any known information about the y_js, both valid constraints and "don't be stupid" constraints that exclude feasible but no optimal solutions. Such constraints can often be deduced from a practical understanding of the problem domain.

6   Branch and Bound with Convex Quasi-Relaxations

In the methods presented so far, the discrete variables need have no particular domain. However, in many applications the discrete variables are real-valued, as for example when they are discretized continuous variables. In such cases it may be advantageous to have a relaxation in both the x and y variables, so that one can branch on y_j's by splitting intervals. The solution of the relaxation would indicate where to split. Thus for example if $y_j \in [y_j^L, y_j^U]$ and the solution value of y_j in the relaxation lies between discrete values $v, v' \in Y_j$, one would split the interval into $[y_j^L, v]$ and $[v', y_j^U]$. The relaxation may therefore accelerate the search not only by providing bounds, but by providing split points that lead more quickly to feasible solutions.

This strategy is practical, however, only when a *convex* relaxation involving the y variables is available. Such a relaxation normally cannot be obtained by relaxing y_j's domain Y_j to a continuous interval, since the resulting problem is in general nonconvex.

Even when a convex relaxation is unavailable, however, it may be possible to construct a convex *quasi-relaxation* that is equally useful for obtaining lower bounds. A quasi-relaxation of a problem $\min\{f(x) \mid x \in S\}$ is a problem $\min\{f'(x) \mid x \in S'\}$ with the property that for any $x \in S$, there exists an $x' \in S'$ for which $f(x') \leq f(x)$. It is clear that the optimal value of the quasi-relaxation, if it exists, provides a valid lower bound on the optimal value of the original problem.

The following theorem provides conditions under which one may construct a convex quasi-relaxation for problem (1). Let function $g(x, y_j)$ be convex in x when $g(x, v)$ is convex for any $v \in Y_j$. Also let $g(x, y_j)$ be *semihomogeneous* in x if

$$g(\alpha x, v) \leq \alpha g(x, v) \text{ for all } \alpha \in [0, 1], x \in \mathbb{R}^n, v \in Y_j \quad (a)$$
$$g(0, y_j) = 0 \text{ for all } y_j \in Y_j \quad (b)$$

$$(11)$$

Theorem 2. *Suppose each $g_i^j(x, y_j)$ in (1) is convex in x and satisfies at least one of the following conditions:*

1. $g_i^j(x, y_j)$ is convex.
2. $g_i^j(x, y_j)$ is semihomogeneous in x and concave in y_j.

Let (i, j) belong to J_1 when g_i^j satisfies condition 1 and J_2 otherwise. Suppose also that $x^L \leq x \leq x^U$ and $y^L \leq y \leq y^U$. Then the following is a convex quasi-relaxation of (1):

$$\begin{aligned}
&\text{minimize } x_1 \\
&\text{subject to } g_i^j\left(x, \alpha_j y_j^L + (1 - \alpha_j) y_j^U)\right) \leq 0, \quad \text{all } (i, j) \in J_1 \quad (a) \\
&\qquad\qquad g_i^j(x^{j1}, y_j^L) + g_j^i(x^{j2}, y_j^U) \leq 0, \quad \text{all } (i, j) \in J_2 \quad (b) \\
&\qquad\qquad \alpha_j x^L \leq x^{j1} \leq \alpha_j x^U, \quad \text{all } j \in J \quad (c) \\
&\qquad\qquad (1 - \alpha_j) x^L \leq x^{j2} \leq (1 - \alpha_j) x^U \quad \text{all } j \in J \quad (d) \\
&\qquad\qquad x = x^{j1} + x^{j2}, \quad \text{all } j \in J \quad (e) \\
&\qquad\qquad x^{j1}, x^{j2} \in \mathbb{R}^n, \quad \alpha_j \in [0, 1], \quad \text{all } j \in J
\end{aligned} \qquad (12)$$

Furthermore, if each α_j is 0 or 1 in the optimal solution of (12), then (12) has the same optimal value as (1).

Proof. We first observe that (12) is convex. Constraint (a) is convex because $g_i^j(x, y_j)$ is convex for $(i, j) \in J_1$, and a convex function composed with an affine function is convex. Constraint (b) is convex because $g_i^j(x, y_j)$ is convex when y_j is fixed. The remaining constraints are linear.

To show that (12) is a quasi-relaxation, take any feasible solution $(\bar{x}, \bar{y})$ of (1) and construct a feasible solution for (12) as follows. For each $j \in J$ choose $\alpha_j \in [0, 1]$ so that $\bar{y}_j = \alpha_j y_j^L + (1 - \alpha_j) y_j^U$. Set $x^{j1} = \alpha_j \bar{x}$, $x^{j2} = (1 - \alpha_j) \bar{x}$, and $x = x^{j1} + x^{j2}$. To see that this produces a feasible solution of (12), note first that constraints (a) and (c)-(e) are satisfied by construction. Constraint (b) is also satisfied, since for $(i, j) \in J_2$ we have

$$\begin{aligned}
g_i^j(x^{j1}, y_j^L) + g_i^j(x^{j2}, y_j^U) &= g_i^j\left(\alpha_j \bar{x}, y_j^L\right) + g_i^j\left((1 - \alpha_j)\bar{x}, y_j^U\right) \\
&\leq \alpha_j g_i^j\left(\bar{x}, y_j^L\right) + (1 - \alpha_j) g_i^j\left(\bar{x}, y_j^U\right) \leq g_i^j\left(\bar{x}, \alpha_j y_j^L\right) + g_i^j\left(\bar{x}, (1 - \alpha_j) y_j^U\right) \\
&= g_i^j(\bar{x}, \bar{y}_j) \leq 0
\end{aligned}$$

where the first inequality is due to the semihomogeneity of $g_i^j(x, y_j)$ in x, the second to the concavity of $g_i^j(x, y_j)$ in y_j, and the third to the feasibility of $(\bar{x}, \bar{y}_j)$ in (1). Also the objective function value of (12) is less than or equal to (in fact equal to) that of (1), since $x_1 = \bar{x}_1$. Thus (12) is a convex quasi-relaxation of (1).

Finally, when $\alpha_j = 1$ we have $x^{j1} = x$ and $x^{j2} = 0$, and similarly if $\alpha_j = 0$. It easy to verify, using the semihomogeneity of $g_i^j(x, y_j)$ in x, that (12) reduces to (1) when each $\alpha_j \in \{0, 1\}$ and therefore has the same optimal value. This completes the proof.

Let $g(x, y_j)$ be homogeneous in x when $g(\alpha x, y_j) = \alpha g(x, y_j)$ for all $\alpha \geq 0, y_j \in Y_j$.

Corollary 1. *Theorem 2 holds in particular when each $g_i^j(x, y_j)$ is either (a) convex or (b) homogeneous in x and concave in y_j.*

If the global optimization problem (1) satisfies the conditions of Theorem 2, it can be solved by branch and bound as follows. Each node of the search tree is processed as in the algorithm below, where z^U is the value of the best feasible solution found so far (initially $z^U = \infty$), and $[y_j^L, y_j^U]$ is the interval in which y_j is currently constrained to lie (where $y_j^L, y_j^U \in Y_j$). Initially the only unprocessed node is the root node, which is processed first.

1. Compute an optimal solution $\bar{x}, \bar{x}^{j1}, \bar{x}^{j2}, \bar{\alpha}_j$ (for $j \in J$) of the convex quasi-relaxation (12) at the current node. Set $\bar{y}_j = \bar{\alpha}_j y_j^L + (1 - \bar{\alpha}_j) y_j^U$.
2. If $\bar{x}_1 \geq z^U$, go to an unprocessed node and begin with step 1.
3. If some $\bar{\alpha}_j \notin \{0, 1\}$, let v, v' be the values in $Y_j \cap [y_j^L, y_j^U]$ on either side of $\bar{y}_j$ that are closest to $\bar{y}_j$. (Possibly v or v' is identical to $\bar{y}_j$.) Branch on y_j by creating an unprocessed node at which $y_j \in [y_j^L, v]$ and a second unprocessed node at which $y_j \in [v', y_j^U]$. Go to an unprocessed node and begin with step 1.
4. The solution $(\bar{x}, \bar{y})$ is feasible in (1). Set $z^U = \min\{\bar{x}_1, z^U\}$. Go to an unprocessed node and start with step 1.
5. The solution $(\bar{x}, \bar{y})$ is feasible in (1). Set $z^U = \min\{\bar{x}_1, z^U\}$. Go to an unprocessed node and start with step 1.

The algorithm terminates when no unprocessed nodes remain. To ensure termination, one should fix α_j at 0 or 1 (either yields the same result) whenever $y_j^L = y_j^U$.

7 Truss Structure Design

We conclude with a truss structure design problem and show how to solve it with disjunctive programming as well as BBCQ. The model presented here is a simplified version of that described in [1].

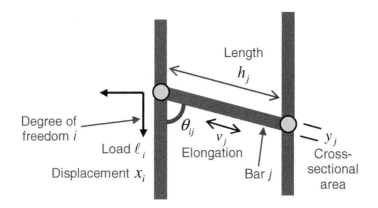

Fig. 1. Notation for a truss structure design problem

The notation is illustrated in Fig. 1. A truss structure consists of a number of bars j joined at nodes, each bar having length h_j and a cost of c_j per unit volume. Each node can move in a specified number of directions. Thus if the problem is solved in three dimensions, there are at most three degrees of freedom at each node. Each degree of freedom i is associated with a load ℓ_i. The decision variables are the thickness (cross-sectional area) y_j of the bars. Other variables are the elongation s_j of bar j, the tension (pulling force) f_j on bar j, and the displacement x_i along degree of freedom i. The objective is to minimize the cost of the bars subject to bounds on elongation and displacement. Stress bounds also exist and are factored into the elongation bounds. The model is

$$
\begin{aligned}
&\text{minimize} \quad \sum_j c_j h_j y_j && \text{cost of bars} \\
&\text{subject to} \quad \frac{E_j}{h_j} y_j s_j = f_j, \ \text{all } j && \text{Hooke's law} \\
&\qquad\qquad \sum_j f_j \cos \theta_{ij} = \ell_i, \ \text{all } i && \text{equilibrium equations} \\
&\qquad\qquad \sum_i x_i \cos \theta_{ij} = s_j, \ \text{all } j && \text{compatibility equations} \\
&\qquad\qquad s_j^L \le s_j \le s_j^U, \ \text{all } j && \text{elongation bounds} \\
&\qquad\qquad x_j^L \le x_j \le x_j^U, \ \text{all } j && \text{displacement bounds} \\
&\qquad\qquad y_j \in Y_j, \ \text{all } j && \text{discrete thicknesses}
\end{aligned}
\tag{13}
$$

where E_j in Hooke's law is the modulus of elasticity for bar j. Since structural bars are generally available only in certain thicknesses, the variables y_j can be regarded as discrete.

Since the problem becomes convex (in fact, linear) when variables y_j are fixed, it is amenable to the methods described above. We will apply disjunctive programming and BBCQ.

First we develop the disjunctive programming approach, using convex hull relaxations. A disjunctive representation of (13) is

$$
\begin{aligned}
&\text{minimize} \quad \sum_j z_j && \text{cost of bars} \\
&\text{subject to} \quad \bigvee_{v \in Y_j}
\begin{bmatrix}
y_j = v \\
z_j \ge c_j h_j v \\
\frac{E_j}{h_j} v s_j = f_j
\end{bmatrix}, \ \text{all } j && \text{cost, Hooke's law} \\
&\qquad\qquad \sum_j f_j \cos \theta_{ij} = \ell_i, \ \text{all } i && \text{equilibrium equations} \\
&\qquad\qquad \sum_i x_i \cos \theta_{ij} = s_j, \ \text{all } j && \text{compatibility equations} \\
&\qquad\qquad s_j^L \le s_j \le s_j^U, \ \text{all } j && \text{elongation bounds} \\
&\qquad\qquad x_j^L \le x_j \le x_j^U, \ \text{all } j && \text{displacement bounds}
\end{aligned}
\tag{14}
$$

Using convex hull relaxations of the disjunctions, we obtain the following convex relaxation of (14):

$$\text{minimize} \quad \sum_j z_j$$

$$\text{subject to } z_j = \sum_{v \in Y_j} z_{jv}, \quad s_j = \sum_{v \in Y_j} s_{jv}, \quad f_j = \sum_{v \in Y_j} f_{jv}, \text{ all } j$$

$$z_{jk} \geq c_j h_j v \beta_{jv}, \text{ all } v \in Y_j, \text{ all } j$$

$$\frac{E_j}{h_j} v s_{jv} = f_{jv}, \text{ all } v \in Y_j, \text{ all } j$$

$$\sum_j f_j \cos \theta_{ij} = \ell_i, \text{ all } i \tag{15}$$

$$\sum_i x_i \cos \theta_{ij} = s_j, \text{ all } j$$

$$\beta_{jv} s_j^L \leq s_{jv} \leq \beta_{jv} s_j^U, \text{ all } j$$
$$x_j^L \leq x_j \leq x_j^U, \text{ all } j$$

$$\sum_{v \in Y_j} \beta_{jv} = 1, \quad \beta_{jv} \geq 0 \text{ all } v \in Y_j, \text{ all } j$$

The relaxation can be simplified, in part by summing each instance of Hooke's law over all $v \in Y_j$.

$$\text{minimize} \quad \sum_j \sum_{v \in Y_j} c_j h_j v \beta_{jv}$$

$$\text{subject to } \frac{E_j}{h_j} \sum_{v \in Y_j} v s_{jv} = f_j, \text{ all } j$$

$$\sum_j f_j \cos \theta_{ij} = \ell_i, \text{ all } i$$

$$\sum_i x_i \cos \theta_{ij} = s_j, \text{ all } j \tag{16}$$

$$\beta_{jv} s_j^L \leq s_{jv} \leq \beta_{jv} s_j^U, \text{ all } j$$
$$x_j^L \leq x_j \leq x_j^U, \text{ all } j$$

$$\sum_{v \in Y_j} \beta_{jv} = 1, \quad \beta_{jv} \geq 0 \text{ all } v \in Y_j, \text{ all } j$$

The disjunctive problem (14) can be solved as an MILP by solving (16) with the integrality condition $\beta_{jv} \in \{0, 1\}$. This MINLP model was in fact proposed by Ghattas, Voudouris and Grossmann [5, 6].

We now develop a BBCQ approach to solving (14). Note first that the model (13) satisfies the conditions of Theorem 2, since all of the constraints are convex (in fact, linear) except Hooke's law, which is convex (in fact, linear) when the y_js are fixed. In addition, the constraint function in Hooke's law is homogeneous in the continuous variables s_j, f_j and concave (in fact, linear) in the discrete variable y_j. The convex quasi-relaxation (12) therefore becomes

$$\text{minimize} \quad \sum_j c_j h_j y_j$$

$$\text{subject to} \quad \frac{E_j}{h_j}\left(y_j^L s_{j1} + y_j^U s_{j2}\right) = f_j, \quad \text{all } j$$

$$\sum_j f_j \cos\theta_{ij} = \ell_i, \quad \text{all } i$$

$$\sum_i x_i \cos\theta_{ij} = s_j, \quad \text{all } j \qquad (17)$$

$$\alpha_j s_j^L \le s_{j1} \le \alpha_j s_j^U, \quad \text{all } j$$
$$(1-\alpha_j)s_j^L \le s_{j2} \le (1-\alpha_j)s_j^U, \quad \text{all } j$$
$$x_j^L \le x_j \le x_j^U, \quad \text{all } j$$
$$x_j = x_{j1} + x_{j2}, \quad \text{all } j$$
$$y_j = \alpha_j y_j^L + (1-\alpha_j)y_j^U, \quad \text{all } j$$
$$\alpha_j \in [0,1], \quad \text{all } j$$

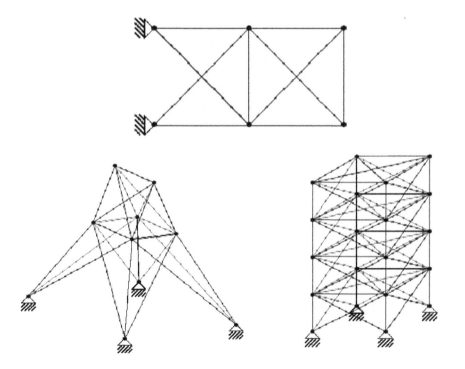

Fig. 2. A 10-bar cantilever truss, 25-bar electrical transmission tower, and 72-bar building

Table 1. Summary of solution times in seconds for MILP and BBCQ applied to truss structure design problems. When there two "loads" (i.e., two sets of loads applied to each degree of freedom), the structure is required to withstand each of the two loads, and a constraint set is written for each one. BBCQ was enhanced with some simple cutting planes when solving the cantilever and tower problems

Problem Instance		MILP	BBCQ
10-bar	1 load	1.3	0.3
cantilever	1 load, wider stress bounds	1.6	0.3
truss	1 load, still wider stress bounds	2.6	1.2
	1 load, still wider stress bounds	2.6	1.4
	2 loads	23.6	5.8
	1 load, displacement bounds	1089.4	67.5
	2 loads, displacement bounds	13743.9	1654.0
25-bar	2 loads	271.7	225.8
transmission			
tower			
Building	72 bars, 2 loads	12692.7	207.9
	90 bars, 2 loads	*	168.9
	108 bars, 2 loads	*	329.4

*No solution after 20 hours (72,000 seconds).

Bollapragada et al. [1] applied both the MILP and BBQC methods to the structural design problems illustrated in Fig. 2. Each structural bar had 11 possible thicknesses. Symmetries in the transmission tower and buildings were exploited to reduce the number of variables. Computational results are summarized in Table 1. MILP was implemented in CPLEX, and BBCQ in C with calls to the CPLEX linear programming solver. All problems were solved on a Sun Sparc Ultra work station.

These results suggest that BBCQ can carry a substantial advantage over a disjunctive approach when the constraint functions satisfy the conditions of Theorem 2.

References

1. S. Bollapragada, O. Ghattas and J. N. Hooker, Optimal design of truss structures by mixed logical and linear programming, *Operations Research* **49** (2001) 42-51.
2. S. Ceria and J. Soares, Convex programming for disjunctive convex optimization, *Mathematical Programming A* **86** (1999) 595–614.
3. V. Chandru and J. N. Hooker, *Optimization Methods for Logical Inference*, John Wiley & Sons (New York, 1999).
4. M. A. Duran and I. E. Grossmann, An outer-approximation algorithm for a class of mixed-integer nonlinear programs, *Mathematical Programming* **36** (1986) 307.
5. O. Ghattas and I. E. Grossmann, MINLP and MILP strategies for discrete sizing structural optimization problems, *Proceedings of the ASCE 10th Conference on Electronic Communication*, Indianapolis (1991).

6. I. E. Grossmann, V. T. Voudouris, and O. Ghattas, Mixed-integer linear programming formulations of some nonlinear discrete design optimization problems, in C. A. Floudas and P. M. Pardalos, eds., *Recent Advances in Global Optimization*, Princeton University Press (1992).

7. I. E. Grossmann and S. Lee, Generalized disjunctive programming: Nonlinear convex hull relaxation, Carnegie Mellon University (2001) submitted.

8. J. Hiriart-Urruty and C. Lemaréchal, *Convex Analysis and Minimization Algorithms*, Vol. 1 (Springer-Verlag, 1993).

9. J. N. Hooker, *Logic-Based Methods for Optimization: Combining Optimization and Constraint Satisfaction*, John Wiley & Sons (2000).

10. J. N. Hooker, Logic-based Benders decomposition for planning and scheduling, manuscript, GSIA, Carnegie Mellon University 2003).

11. J. N. Hooker and M. A. Osorio, Mixed logical/linear programming, *Discrete Applied Mathematics* **96-97** (1999) 395–442.

12. J. N. Hooker and G. Ottosson, Logic-based Benders decomposition, *Mathematical Programming* **96** (2003) 33–60.

13. Jain, V., and I. E. Grossmann, Algorithms for hybrid MILP/CP models for a class of optimization problems, *INFORMS Jorunal on Computing* **13** (2001) 258–276.

14. R. Stubbs and S. Mehrotra, A branch-and-cut method for 0-1 mixed convex programming, *Mathematical Programming* **86** (1999) 515–532.

15. M. Türkay and I. E. Grossmann, Logic-based outer-approximation algorithm for MINLP optimization of process flowsheets, *Computers and Chemical Engineering* **19** (1996) S131–S136.

Accelerating Consistency Techniques and Prony's Method for Reliable Parameter Estimation of Exponential Sums[*]

Jürgen Garloff[1], Laurent Granvilliers[2], and Andrew P. Smith[1]

[1] University of Applied Sciences / FH Konstanz,
Postfach 100543, D-78405 Konstanz, Germany
{garloff, smith}@fh-konstanz.de
[2] LINA, University of Nantes,
BP 92208, F-44322 Nantes cedex 3, France
granvilliers@lina.univ-nantes.fr

Abstract. In this paper the problem of parameter estimation for exponential sums is considered, i.e., of finding the set of parameters (amplitudes as well as decay constants) such that the exponential sum attains values in specified intervals at prescribed time data points. These intervals represent uncertainties in the measurements. An interval variant of Prony's method is given by which a box can be found containing all the consistent values of the parameters. Subsequently this box is tightened by the use of consistency techniques, which are accelerated by the introduction of redundant constraints. The use of interval arithmetic results in enclosures for the consistent values of the parameters which can be guaranteed also in the presence of rounding errors.

Keywords: Parameter estimation, exponential sum, Prony's method, interval arithmetic, constraint propagation, redundant constraint.

1 Introduction

The simulation of complex systems for a wide range of applications dates back to the early development of modern computers. Once a mathematical model is known, the system behaviour can be analysed without the need for practical experimentation. This approach is specifically useful to compute information which cannot easily be obtained in practice or to test extreme situations. It also becomes possible to predict the system behaviour or to optimize system components. In the following, we will consider a family of dynamical systems modeled by the function

$$y(t) = f(\mathrm{x}, t), \tag{1}$$

---
[*] This work has been supported by a PROCOPE project funded by the French Ministry of Education, the French Ministry of Foreign Affairs, the German Academic Exchange Service (DAAD) and the Ministry of Education and Research of the Federal Republic of Germany under contract no. 1705803.

C. Jermann et al. (Eds.): COCOS 2003, LNCS 3478, pp. 31–45, 2005.

where t represents time, and $x \in \mathbf{R}^n$ is the vector of parameters. Each individual system leads to the problem of finding consistent values of parameters.

Let observations of the system be given, that is a series of data $(\tilde{y}_i, t_i)$, $i = 1, \ldots, m$, where $\tilde{y}_i$ is the system output at time t_i. The model-driven inverse problem (*parameter estimation problem*) consists of finding values of x such that the following equations hold:

$$\tilde{y}_i = f(x, t_i), \quad i = 1, \ldots, m.$$

Unfortunately, this problem generally has no solution, since output values may be imprecise and uncertain. Therefore one tries to determine values of the model parameters that provide the best fit to the observed data, generally based on some type of maximum likelihood criterion, which results in minimizing the function

$$\sum_{i=1}^{m} w_i (f(x, t_i) - \tilde{y}_i)^2. \tag{2}$$

It is not uncommon for the objective function (2) to have multiple local optima in the area of interest. However, the standard methods used to solve this problem are local methods that offer no guarantee that the global optimum, and thus the best set of model parameters, has been found. In contrast, methods from global optimization [10, 11, 13] are capable of localizing the global optimum of (2). However, this approach does not take into account that the observed data are affected by uncertainty. Therefore the resulting models may be inconsistent with error bounds on the data.

To take uncertainty into account, we assume that the observed data are corrupted by errors, e.g. measurement errors, $\pm\varepsilon_i, \varepsilon_i \geq 0, i = 1, \ldots, m$. Then the correct value $y_i = f(x^*, \varepsilon_i)$ is within the interval $[\tilde{y}_i - \varepsilon_i, \tilde{y}_i + \varepsilon_i]$, $i = 1, \ldots, m$. More generally, we suppose that y_i is known to be contained in the interval $[a_i, b_i]$. The data driven inverse problem (*parameter set estimation problem*) consists of finding values of x subject to the following system of inequalities:

$$a_i \leqslant f(x, t_i) \leqslant b_i, \quad i = 1, \ldots, m. \tag{3}$$

The aim is to compute a representation of the set Ω of the consistent values of the parameters that may help in decision making. Interval arithmetic and inclusion functions for the model functions are used in [17, 21] to find boxes generated by bisection which are contained in Ω; the union of these boxes constitutes an inner approximation of Ω. Also, boxes are identified which contain part of the boundary of Ω or contain only inconsistent values; boxes of this second category can be used to construct an outer approximation of the set of inconsistent values. However, such an approach can not handle large initial boxes or problems with many parameters. Therefore, interval constraint propagation techniques are introduced in [19] to drastically reduce the number of bisections.

In this paper, we concentrate on models of exponential sums arising in many applications such as, e.g., pharmacokinetics [14, 26]. It is well-known, e.g. [5], p. 242, and [22], that parameter estimation of exponential sums is notoriously sensitive to data perturbations. Two complementary techniques are applied. The first one is an interval variant of Prony's method [22, 25], which aims to compute an initial domain for the parameters to be estimated. The second one is applied after problem (3) is transformed into a set of equalities and is the symbolic generation of redundant constraints in order to accelerate constraint propagation. The challenge is to compute constraints leading to more precision in the numerical process, to control the amount of symbolic computations and to limit the number of redundancies in order to avoid slow-downs of the whole solving procedure.

The outline of this paper is as follows. The basics of interval arithmetic and constraint satisfaction techniques are presented in Section 2. The new methods are introduced in Section 3. A numerical example is given in Section 4. We finally conclude in Section 5.

2   Preliminaries

2.1   Interval Arithmetic

We consider the following sets: the set $\mathbf{R}$ of real numbers including the infinities, the finite set $\mathbf{F}$ of floating point numbers and the finite set $\mathbf{I}$ of closed intervals spanned by two floating point numbers. Every interval $\mathbf{x} \in \mathbf{I}$ is denoted by $[\underline{x}, \overline{x}]$ and is defined as the set of real numbers $\{x \in \mathbf{R} \mid \underline{x} \leqslant x \leqslant \overline{x}\}$.

Interval arithmetic [23] is a set theoretic extension of real arithmetic. The operations are implemented by floating-point computations with interval bounds according to monotonicity properties. For instance, the sum $[a, b] + [c, d]$ is equal to $[a + c, b + d]$, provided that the left bound is downward rounded and the right bound is upward rounded. Interval reasonings can be extended to complex functions using the so-called interval evaluation method. Given a function $f : \mathbf{R}^n \to \mathbf{R}$, let each real number in the expression of f be replaced by the interval spanned by floating point numbers obtained by rounding this real number downward and upward, each variable be replaced with its domain, and each operation be replaced with the corresponding interval operation. Then the interval expression can be evaluated using interval arithmetic, which results in a superset of the range of f over the domain of the variables.

2.2   Consistency Techniques

A numerical constraint satisfaction problem (NCSP) is given by a set of variables $\{x_1, \ldots, x_n\}$, each variable x_i lying in an interval domain $\mathbf{x}_i$, and a set of constraints over the real numbers $\{c_1, \ldots, c_m\}$. The solution set of a NCSP is defined as the set

$$\{a \in \mathbf{R}^n \mid c_1(a) \wedge \cdots \wedge c_m(a)\},$$

where each constraint c_j is considered as a relation.

Consistency techniques aim to reduce the Cartesian product of variables domains $\mathbf{x}_1 \times \cdots \times \mathbf{x}_n$, which defines the search space called a box. Most of the reduction algorithms are based on constraint projections. The projection of a constraint $c(x_1, \ldots, x_n)$ over a variable x_i is the set

$$\Pi_i(c) = \{a_i \in \mathbf{x}_i \mid \forall j \in \{1, \ldots, n\} \setminus \{i\}, \exists a_j \in \mathbf{x}_j : c(a_1, \ldots, a_n)\}.$$

It follows that the reduction step

$$\mathbf{x}_i := \Pi_i(c)$$

is reliable since each value belonging to the complementary set cannot be extended in a solution of the NCSP. In practice projections are reliably approximated by means of interval computations. For example the inversion algorithm uses the so-called relational interval arithmetic [8]. A numerical inversion procedure has been described as a chain rule in [16].

Example 1. Consider the constraint $2x_1 - x_2^2 = 4$, given $(x_1, x_2) \in [-3, 3] \times [1, 3]$. The computation of its projection over x_1 by the chain rule can be explained as follows. Define an equivalent constraint, where the left-hand term is reduced to x_1, namely $x_1 = (4 + x_2^2) \div 2$. Evaluate the right-hand term using interval arithmetic. The interval $[2.5, 6.5]$ is computed, and it is intersected with the domain of x_1. The new domain of x_1 is equal to $[2.5, 3]$. Thus, the set of values $[-3, 2.5)$ has been shown to be locally inconsistent with the given constraint.

Given a set of constraints, constraint projections have to be processed in sequence in order to obtain the consistency of the whole problem. The corresponding iterative algorithm is called constraint propagation. The result is a new box that contains the solution set. In order to separate the solutions, constraint propagation has to be embedded in a more general bisection algorithm. Boxes are reduced and then bisected until every box is sufficiently small.

2.3   Data Fitting Problems as NCSPs

Problem (3) should be transformed before propagation for two reasons. First, the variable y_i has to be explicit in order to reduce the error bounds. Second, each data value leads to two inequalities involving the term $f(\mathbf{x}, t_i)$. Since constraints are processed independently, an efficient approach consists of sharing computations over this term. Problem (3) is equivalent to the following set of existentially quantified equations

$$\exists y_i \in [a_i, b_i] : \; y_i = f(\mathbf{x}, t_i), \quad i = 1, \ldots, m. \tag{4}$$

Now, quantifiers can be removed, making the variables y_i first-class variables. This leads to Problem (5):

$$y_i = f(\mathbf{x}, t_i), \quad i = 1, \ldots, m. \tag{5}$$

Problems (4) and (5) are equivalent for computations of projections over the parameters. In fact, quantifiers just introduce an intermediary level of projections, which is of no benefit. It can clearly be seen that constraint propagation for Problem (5) is on average twice as fast as propagation for Problem (3).

2.4   Exponential Sums

We consider now a model with exponential sums, as follows:

$$f(\mathbf{x}, t) = \sum_{j=1}^{p} x_{2j-1} \exp(-x_{2j}t), \quad n = 2p. \tag{6}$$

In fact three problems occur when exponential sums are processed by consistency techniques. The first problem is the evaluation of the exponential function over positive real numbers far from 0. For instance consider a term $\exp(-tx)$ given $t = 100$ and suppose that x is negative. If x is smaller than -8 then $\exp(-tx)$ is evaluated to $+\infty$ on a 64-bit machine. In this case, interval-based methods are powerless. This weakness points out the needs for getting an *a priori* tight search space of parameters.

The second problem concerns slow convergences in constraint propagation. The cause is that two exponential sums from two different constraints have a similar shape. For instance consider the terms $f_1(x) = 0.2e^{0.3x} + 1$ and $f_2(x) = 0.5e^{0.4x}$, depicted in Figure 1 (f_2 has the largest slope). Domain reductions are numbered from 1. The first reduction concerns the upper bound of y using f_1. The eliminated box contains no solution of equation $y = f_1(x)$, i.e., no point of the curve of f_1. Then, the upper bound of x is contracted using f_2, and so on. A similar process leads to the reduction of the other bounds. In this case, the number of constraint processing steps using the chain rule is equal to 82.

In practice, the only difference is that the variables are not reduced to real values, but that they belong to real intervals. The intersection of curves becomes an intersection of surfaces. In this case, inefficiencies of constraint propagation remain.

The third problem is inherent to local approaches, since a sequence of local reasonings may not derive global information. Many techniques try to overcome this problem, one of them being the use of redundant constraints in the constraint propagation algorithm. A constraint is said to be *redundant* with respect to a

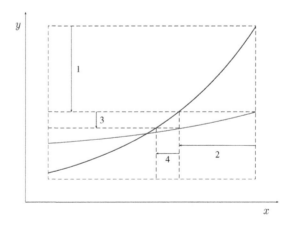

Fig. 1. Constraint Propagation over Two Exponential Terms

set of constraints if it does not influence the solution set. Redundant constraints can be derived from the set using combination and simplification procedures, for instance Gröbner basis techniques for polynomials [7]. The interesting feature is that combination is a means for sharing information between constraints. The main challenge is to control the amount of symbolic computations, to compute constraints able to improve the precision of consistency techniques, and to limit the number of redundant constraints in order avoid slow-downs in constraint propagation.

3 Acceleration Methods

3.1 Prony's Method

Given the model (6), we wish to find *decay constants* x_{2j} and *amplitudes* x_{2j-1}, $j = 1, \ldots, p$, such that (5) is satisfied at equidistant $t_i = t_0 + ih$, $i = 1, \ldots, m$, with given stepsize h. A method to accomplish this task is Prony's method [25], cf. Chap. IV, §23 of [22], which dates back to the 18th century. This method relies on the observation that a function of the form (6) satisfies a linear difference equation with constant coefficients. We concentrate here on the case $p = 2$. We choose a fixed group of four time data points, selected from the set $\{1, \ldots, m\}$, say $\{1, 2, 3, 4\}$. Prony's method then first requires the solution of the following system of two linear equations in the unknowns ζ_1 and ζ_2.

$$\begin{pmatrix} y_1 & y_2 \\ y_2 & y_3 \end{pmatrix} \begin{pmatrix} \zeta_1 \\ \zeta_2 \end{pmatrix} = - \begin{pmatrix} y_3 \\ y_4 \end{pmatrix}. \tag{7}$$

The solution (ζ_1, ζ_2) of this system provides the coefficients of a quadratic

$$q(u) = u^2 + \zeta_2 u + \zeta_1. \tag{8}$$

If the zeros u_1 and u_2 of q are distinct and positive then the decay constants are given by $\{x_2, x_4\} = \{\log(u_1)/h, \log(u_2)/h\}$. Finally, we obtain the amplitudes x_1 and x_3 from the solution of a second system of two linear equations

$$\begin{pmatrix} 1 & 1 \\ u_1 & u_2 \end{pmatrix} \begin{pmatrix} z_1 \\ z_3 \end{pmatrix} = \begin{pmatrix} y_1 \\ y_2 \end{pmatrix} \tag{9}$$

with $x_k = e^{-t_1 x_{k+1}} z_k$, $k = 1, 3$.

Now consider the interval problem (4). We want to find intervals $\mathbf{x}_1, \ldots, \mathbf{x}_4$, such that all $x_j \in \mathbf{x}_j$, $j = 1, \ldots, 4$, for which

$$f(\mathbf{x}, t_i) \in [a_i, b_i], \quad i = 1, \ldots, m. \tag{10}$$

By changing to the interval data given by (4), Prony's method now requires the solution of interval variants of the two linear systems (7) and (9) and the enclosure of the zero sets of the interval polynomial corresponding to (8) [1]. Special

---

[1] A preliminary version of the interval variant of Prony's method was given in Sect. 5.2 of [12].

care has to be taken to find tight intervals for the decay constants and amplitudes. To determine enclosures for the zero sets of q in the case that the roots are positive and can be separated, we compute an enclosure for the largest positive root by the well-known formula and a respective enclosure for the smallest positive root by an interval variant of Vietà's method.

For a system of p linear interval equations in p unknowns

$$[A]x = [b] \tag{11}$$

the (general) solution set is defined as the set

$$\Sigma = \{x \in \mathbf{R}^p \mid \exists A \in [A], b \in [b] \; : \; Ax = b\}. \tag{12}$$

Here we assume that the interval matrix is nonsingular, i.e., it contains only nonsingular real matrices. We are interested in the hull of the solution set, i.e., the smallest axis aligned box containing Σ.

For the system of two linear interval equations corresponding to (9), we can easily compute the hull of the solution set by the method presented in [3], cf. [24] p. 97. The system (7) exhibits two dependencies: The system matrix is symmetric and the coefficient in its bottom right corner is equal to the negation of the first entry of the right hand side. So it is natural to consider in the interval problem the symmetric solution set Σ_{sym} [1, 2], [24], Sect. 3.4, which is the solution set restricted to the systems with symmetric matrices, and the even smaller solution set, denoted by Σ^*_{sym}, obtained when in addition the dependency on the first entry of the right hand side is taken into account. With elementary computations (which are delegated to the Appendix) it is possible to determine the hulls of these structured solution sets. In Figure 2, these three solution sets together with their hulls for the following system

$$\begin{pmatrix} [1,3] & [0,1] \\ [0,1] & [-4,-1] \end{pmatrix} \begin{pmatrix} \zeta_1 \\ \zeta_2 \end{pmatrix} = - \begin{pmatrix} [-4,-1] \\ [-1,2] \end{pmatrix} \tag{13}$$

are displayed. The general solution set Σ consists of the whole shaded region and the symmetric solution set Σ_{sym} consists of the regions shaded in medium and dark grey. The dark grey region is the solution set Σ^*_{sym}. At least the first two solution sets can be obtained by analytical methods, cf. [1, 2], but are determined here by the computation of the solutions of a large number of real systems corresponding to boundary and interior points of the interval matrix and the interval right hand side.

If the interval system corresponding to (7) is singular, one should check whether the underlying problem is not better described by a single exponential term, i.e., we have $p = 1$ in (6). In fact, if

$$\tilde{y}_i := x_1 \exp(-x_2(t_0 + ih)) \in [a_i, b_i], \quad i = 1, 2, 3, \tag{14}$$

holds true, then it follows that

$$0 = \tilde{y}_1 \tilde{y}_3 - \tilde{y}_2^2. \tag{15}$$

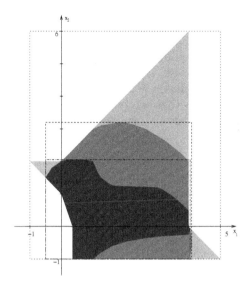

Fig. 2. The three solution sets Σ, Σ_{sym}, and Σ_{sym}^* and their hulls for system (13)

We mention two possibilities for tightening the enclosures for the parameters obtained in this way: we can choose another group of four time data points, compute again enclosures for the parameters and intersect with the enclosures obtained for the first group. Continuing in this way, we successively improve the quality of the enclosures. If an intersection becomes empty, we have then proven that there is no exponential function of the form (6) which solves the real interpolation problem with data taken from the intervals given in (4).

Another possible improvement is obtained as follows: If we plug on the right hand side of (6) the intervals $\mathbf{x}_j$ into x_j, $j = 1, \ldots, 4$, then we will obtain an interval function. If the evaluation of this function at a time data point results in an interval which is not equal to or a superset of the original data interval, we have proven that certain measurements are not possible. If this difference is large, we may conclude that measurements have not been made precisely enough.

A salient feature of the above approach is that if this method works, i.e., the two interval systems are nonsingular and the roots can be separated, we obtain an enclosure for the parameters without any prior information on the decay constants and amplitudes. Such prior information is normally required for the use of interval methods, e.g., [20]. Often one has to choose an unnecessarily wide starting box which is assumed to contain all feasible values of interest. Application of a subdivision method then results in a large number of subdivision steps. Therefore, Prony's method is predestinated to be used as a preprocessing step for more sophisticated methods. The amount of computational effort is negligible.

3.2   Redundant Computations

Several transformation techniques [18] of exponential sums have been proposed, mainly for the case of data equidistant in time, cf. Sect. 3.1. The other situation has been studied less. However, we will see that constraint propagation may be greatly improved if well-chosen redundant constraints are generated. Given Problem (5), the basic idea is that two terms in the same column can be divided to generate a redundant constraint, as follows:

$$\begin{cases} u_{ij} = x_{2j-1} \exp(-x_{2j}t_i), \\ u_{kj} = x_{2j-1} \exp(-x_{2j}t_k), \\ u_{ij} = u_{kj} \exp(x_{2j}(t_k - t_i)). \end{cases} \tag{16}$$

The simplification consists in eliminating variable x_{2j-1} from the last constraint. The system is then rewritten as follows:

$$\begin{cases} y_i = \sum_{j=1}^{p} u_{ij}, & i = 1, \ldots, m, \\ u_{ij} = x_{2j-1} \exp(-x_{2j}t_i), & i = 1, \ldots, m, \ j = 1, \ldots, p, \\ u_{ij} = u_{kj} \exp(x_{2j}(t_k - t_i)), & 1 \leqslant i < k \leqslant m, \ j = 1, \ldots, p. \end{cases} \tag{17}$$

The number of exponential terms in the system potentially grows from mp to $mp + 0.5m(m-1)p$. In fact the complexity is increased by a non-constant factor $O(m)$. Even if the precision of numerical computations is improved by the use of the redundant constraints, too many constraints to be considered during propagation may induce a slow-down. We then show how to keep the same complexity while filtering the necessary constraints. Consider the first three constraints from the initial system c_1, c_2, and c_3, and let j represent the j-th column. The symbolic step is an elimination procedure which combines two constraints in order to remove the variable x_{2j-1}. The aim is to derive a constraint whose projection over x_{2j} can be efficiently computed. As a consequence, a redundancy, e.g., between c_1 and c_2, is equivalent to an existentially quantified formula, as follows:

$$\exists x_{2j-1} \ c_1 \wedge c_2.$$

Now, suppose that the two following redundancies are available:

$$\exists x_{2j-1} \ c_1 \wedge c_2, \qquad \exists x_{2j-1} \ c_2 \wedge c_3.$$

It can be shown that the third redundancy c defined by $\exists x_{2j-1} \ c_1 \wedge c_3$ is useless for reducing the domain of x_{2j}. Suppose that one value of x_{2j} does not allow the satisfaction of c. Then either c_1 or c_3 is violated, and so do the first two redundancies. We then conclude that c is useless. As a consequence, it suffices to consider per column the redundancies between two consecutive rows. The number of redundant constraints is then equal to $(m-1)p$.

Example 2. Consider the following instance of (16), where variables u are computed by simulation, given the parameter values $(10, 0.5)$:

$$\begin{cases} (i, j, k) = (1, 1, 2) \\ (t_1, t_2) = (2, 5) \\ \quad u_{11} = x_1 \exp(-x_2 t_1) \\ \quad u_{21} = x_1 \exp(-x_2 t_2). \end{cases} \tag{18}$$

Now, find $x_1 \in \mathsf{x}_1$ and $x_2 \in \mathsf{x}_2$ such that the equations of (18) are satisfied. First of all, if the domains are such that the exponential terms are evaluated to $+\infty$, e.g., for $\mathsf{x}_1 = \mathsf{x}_2 = [-1000, 1000]$, then consistency techniques are powerless. If the domains are tighter, e.g., $\mathsf{x}_1 = \mathsf{x}_2 = [-100, 100]$, then one box enclosing the solution is derived after 94 calls to the chain rule:

$$[9.9999999963, 10.000000004] \times [0.49999999988, 0.50000000013].$$

The redundant constraint is

$$u_{11}/u_{21} = \exp(x_2(t_2 - t_1)).$$

If it is added to the system, the number of calls decreases to 5.

In fact more work can be done symbolically. Let I, J denote the domains of u_{ij} and u_{kj} and let K denote the domain of x_{2j}. Then a new domain for variable x_{2j} can be computed by the following interval expression:

$$x_{2j} := K \cap \left(\frac{1}{t_k - t_i} \cdot \log \left(\frac{I}{J} \right) \right). \tag{19}$$

4   A Numerical Example

Software. The software RealPaver [15] is used for the tests. Given a model of exponential sums and a series of measurements together with error bounds, the aim is to compute the convex hull of the set of consistent values of the unknowns. In the following, the same tuning of algorithms is used, namely a fixed number of boxes in the bisection process and a fixed maximum computation time. This way, the precision of resulting boxes can be compared for different input systems.

Benchmark. Consider the following problem, consisting of four time-equidistant measurements:

$$\begin{aligned}
x_1 e^{4.387 x_2} + x_3 e^{4.387 x_4} &\in [-0.304, -0.298] \\
x_1 e^{12.069 x_2} + x_3 e^{12.069 x_4} &\in [21.43, 21.86] \\
x_1 e^{19.751 x_2} + x_3 e^{19.751 x_4} &\in [171.9, 175.3] \\
x_1 e^{27.434 x_2} + x_3 e^{27.434 x_4} &\in [1257, 1282]
\end{aligned}$$

Results. Starting with an initial box

$$[-100, 100] \times [-10, 10] \times [-100, 100] \times [-10, 10]$$

RealPaver computes no reduction. If the 6 redundant constraints are used, then the box is reduced to

$$[-100, 100] \times [-1.326, 10] \times [-100, 100] \times [-1.326, 10].$$

There is clearly a need for using Prony's method to obtain a tight initial box. For the considered problem, Prony's method computes (within 0.01s) the following enclosures for the set of parameters:

$$[-6.673, -3.374] \times [-0.130, 0.014] \times [0.911, 1.344] \times [0.247, 0.266].$$

RealPaver then computes the following new box:

$$[-5.881, -3.618] \times [-0.124, 0.014] \times [0.977, 1.256] \times [0.251, 0.262].$$

This precision is improved if the redundant constraints are used, as follows:

$$[-5.872, -3.740] \times [-0.123, 0.014] \times [0.991, 1.223] \times [0.252, 0.262].$$

5   Conclusion

In this paper, we have shown that constraint satisfaction techniques have to be improved in order to process exponential-based models, which are often ill-conditioned. For this purpose, two techniques have been introduced, namely an interval variant of Prony's method and a symbolic procedure. The main goal is to improve the tightness of the bounds for the parameters, whilst keeping the computation time unchanged (or improved).

In a bounded-error context, the problem is to solve a set of inequalities. A powerful approach is to use inner computations to approximate the interior of the solution set, which is often a continuum, using boxes. For this purpose, we believe that three techniques should be combined: inner computations using constraint negations [6], an inner box extension method [9] and interior algorithms based on local search.

A serious limitation of Prony's method is that it requires equidistant time data points. However, many examples in the literature contain at least some equidistant data points. If the measurements provide a group of at least four such points, then we can apply Prony's method as a preprocessing step to deliver a suitable initial box. In a future paper, we will report on Prony's method for functions (6) comprising three exponential terms ($p = 3$).

References

1. G. Alefeld, V. Kreinovich, and G. Mayer. On the shape of the symmetric, persymmetric, and skew-symmetric solution set. *SIAM J. Matrix Anal. Appl.*, 18:693–705, 1997.
2. G. Alefeld and G. Mayer. On the symmetric and unsymmetric solution set of interval systems. *SIAM J. Matrix Anal. Appl.*, 16:1223–1240, 1995.
3. N. Apostolatos and U. Kulisch. Grundzüge einer Intervallrechnung für Matrizen und einige Anwendungen. *Elektron. Rechenanlagen*, 10:73–83, 1968.
4. H. Beeck. Über intervallanalytische Methoden bei linearen Gleichungssystemen mit Intervallkoeffizienten und Zusammenhänge mit der Fehleranalysis. Dissertation, Technical University Munich, 1971.

5. R. Bellmann. *Methods of Nonlinear Analysis*, volume I. Academic Press, New York, London, 1970.
6. F. Benhamou and F. Goualard. Universally quantified interval constraints. In R. Dechter, editor, *Proceedings of International Conference on Principles and Practice of Constraint Programming*, volume 1894 of *Lecture Notes in Computer Science*, pages 67–82, Singapore, 2000. Springer-Verlag.
7. F. Benhamou and L. Granvilliers. Automatic generation of numerical redundancies for nonlinear constraint solving. *Reliable Computing*, 3(3):335–344, 1997.
8. J. G. Cleary. Logical arithmetic. *Future Computing Systems*, 2(2):125–149, 1987.
9. H. Collavizza, F. Delobel, and M. Rueher. Extending consistent domains of numeric CSPs. In T. Dean, editor, *Proceedings of International Joint Conference on Artificial Intelligence*, pages 406–413, Stockholm, Sweden, 1999. Morgan Kaufmann.
10. W. R. Esposito and C. A. Floudas. Global optimization in parameter estimation of nonlinear algebraic models via the error–in–variables approach. *Industrial and Engineering Chemistry Research*, 37:1841–1858, 1998.
11. W. R. Esposito and C. A. Floudas. Parameter estimation of nonlinear algebraic models via global optimization. *Computers and Chemical Engineering*, 22:S213–S220, 1998.
12. J. Garloff. Untersuchungen zur Intervallinterpolation. Dissertation, University of Freiburg, *Freiburger Intervall-Berichte* 80/5, 1980.
13. C.-Y. Gau and M. A. Stadtherr. Nonlinear parameter estimation using interval analysis. *AIChE Symp. Ser.*, 94(304):444–450, 1999.
14. M. Gibaldi and D. Perrier. *Pharmacokinetics*. Marcel Dekker, Inc., New York, 1982.
15. L. Granvilliers. *RealPaver User's Manual, version 0.3*, 2003. Available at www.sciences.univ-nantes.fr/info/perso/permanents/granvil/realpaver.
16. L. Granvilliers and F. Benhamou. Progress in the solving of a circuit design problem. *Journal of Global Optimization*, 20(2):155–168, 2001.
17. E. P. Hofer, B. Tibken, and M. Vlach. Traditional parameter estimation versus estimation of guaranteed parameter sets. In W. Krämer and J. Wolff von Guddenberg, editors, *Scientific Computing, Validated Numerics, Interval Methods*, pages 241–254. Kluwer Academic Publishers, Boston, Dordrecht, London, 2001.
18. K. Holmström and J. Petersson. A review of the parameter estimation problem of fitting positive exponential sums to empirical data. *Applied Mathematics and Computations*, 126:31–61, 2002.
19. L. Jaulin. Interval constraint propagation with application to bounded-error estimation. *Automatica*, 36:1547–1552, 2000.
20. L. Jaulin, M. Kieffer, O. Didrit, and E. Walter. *Applied Interval Analysis*. Springer, London, Berlin, Heidelberg, 2001.
21. L. Jaulin and E. Walter. Set inversion via interval analysis for nonlinear bounded-error estimation. *Automatica*, 29(4):1053–1064, 1993.
22. C. Lanczos. *Applied Analysis*. Prentice-Hall, Englewood Cliffs, NJ, 1956.
23. R. E. Moore. *Interval Analysis*. Prentice-Hall, Englewood Cliffs, NJ, 1966.
24. A. Neumaier. *Interval Methods for Systems of Equations*. Cambridge University Press (Encyclopedia of Mathematics and its Applications), Cambridge, 1990.
25. R. Prony. Essai experimental et analytique sur les lois de la dilatabilité des fluides élastiques et sur celles de la force expansive de la vapeur de l'eau et de la vapeur de l'alkool, à différentes températures. *Journal de l'Ecole Polytechnique*, 1(2):24–76, 1795.
26. M. Weiss. *Theoretische Pharmakokinetik*. Verlag Gesundheit, Berlin, 1990.

Appendix

Determination of the Hulls of the Three Solution Sets of the Linear Interval System Appearing in Prony's Method

It is well-known, e.g. [4], that the hull of the (general) solution set of (11) can be obtained as the hull of the solutions of all the vertex systems of (11), i.e., the systems of real equations with coefficients being identical to endpoints of the respective coefficient intervals.[2] Therefore, in the case $p = 2$ we have to solve 2^6 point systems. Consider now the symmetric system

$$\begin{pmatrix} [\underline{a}_1, \overline{a}_1] & [\underline{a}_2, \overline{a}_2] \\ [\underline{a}_2, \overline{a}_2] & [\underline{a}_3, \overline{a}_3] \end{pmatrix} \begin{pmatrix} x_1 \\ x_2 \end{pmatrix} = \begin{pmatrix} [\underline{b}_1, \overline{b}_1] \\ [\underline{b}_2, \overline{b}_2] \end{pmatrix}$$

and one of its point systems

$$\begin{pmatrix} a_1 & a_2 \\ a_2 & a_3 \end{pmatrix} \begin{pmatrix} x_1 \\ x_2 \end{pmatrix} = \begin{pmatrix} b_1 \\ b_2 \end{pmatrix}.$$

Assume that the matrix is nonsingular. Then it is easy to see that both components of the solution vector (x_1 and x_2) are monotonic with respect to a_1, a_3, b_1, and b_2. Therefore, x_1 and x_2 can attain their minimum and maximum only at the endpoints of the intervals $[a_1]$, $[a_3]$, $[b_1]$, and $[b_2]$. Since

$$\frac{\partial x_1}{\partial a_2} = \frac{-b_2 a_2^2 + 2a_3 b_1 a_2 - a_1 a_3 b_2}{(a_1 a_3 - a_2^2)^2}$$

x_1 can only take its minimum and maximum when $a_2 \in \{\underline{a}_2, \overline{a}_2\}$ or

$$b_2 a_2^2 - 2a_3 b_1 a_2 + a_1 a_3 b_2 = 0. \tag{20}$$

Similarly, x_2 can only take its extreme values when $a_2 \in \{\underline{a}_2, \overline{a}_2\}$ or

$$b_1 a_2^2 - 2a_1 b_2 a_2 + a_1 a_3 b_1 = 0. \tag{21}$$

So we have to solve all possible 2^5 vertex systems. We additionally have to consider point systems generated as follows: For each of the 2^4 possible combinations

$$a_1 \in \{\underline{a}_1, \overline{a}_1\}, a_3 \in \{\underline{a}_3, \overline{a}_3\}, b_1 \in \{\underline{b}_1, \overline{b}_1\}, b_2 \in \{\underline{b}_2, \overline{b}_2\},$$

solve the two quadratic equations (20) and (21); this gives up to four values $a_2^{(i)}$, $i = 1, 2, 3, 4$. Discard any $a_2^{(i)}$ for which $a_2^{(i)} \notin [\underline{a}_2, \overline{a}_2]$. Solve the point systems for the remaining $a_2^{(i)}$. Thus we need to solve at most $4 * 2^4$ extra point systems altogether. After at most 96 point systems are solved, we have to compute the smallest box containing all the solutions (x_1, x_2) generated in this way. This box provides the hull of the symmetric solution set.

---

[2] For a more tractable approach see Chap. 6 in [24].

We consider now the linear interval system

$$\begin{pmatrix} [\underline{a}_1, \overline{a}_1] & [\underline{a}_2, \overline{a}_2] \\ [\underline{a}_2, \overline{a}_2] & [\underline{b}_1, \overline{b}_1] \end{pmatrix} \begin{pmatrix} x_1 \\ x_2 \end{pmatrix} = \begin{pmatrix} [\underline{b}_1, \overline{b}_1] \\ [\underline{b}_2, \overline{b}_2] \end{pmatrix}. \tag{22}$$

This is the same system as before, except that an extra dependency, viz. $a_3 = b_1$ has been introduced. Note that we have suppressed the minus-sign appearing on the right hand side of (7) for simplicity. Affixing a minus-sign on the right hand side of (7) results in a reflection of the solution set at the origin. Consider the point system

$$\begin{pmatrix} a_1 & a_2 \\ a_2 & b_1 \end{pmatrix} \begin{pmatrix} x_1 \\ x_2 \end{pmatrix} = \begin{pmatrix} b_1 \\ b_2 \end{pmatrix}.$$

Again, assume that the matrix is nonsingular. As before, we have that x_1 and x_2 are monotonic with respect to a_1 and b_2. In addition, x_2 is also monotonic with respect to b_1. This leaves

$$\frac{\partial x_1}{\partial a_2} = \frac{-b_2 a_2^2 + 2b_1^2 a_2 - a_1 b_1 b_2}{(a_1 b_1 - a_2^2)^2}, \tag{23}$$

$$\frac{\partial x_1}{\partial b_1} = \frac{a_1 b_1^2 - 2a_2^2 b_1 + a_1 a_2 b_2}{(a_1 b_1 - a_2^2)^2}, \tag{24}$$

$$\frac{\partial x_2}{\partial a_2} = \frac{-b_1 a_2^2 + 2a_1 b_2 a_2 - a_1 b_1^2}{(a_1 b_1 - a_2^2)^2}. \tag{25}$$

We have to solve a number of point systems, which fall into four categories (see below). After these point systems are solved, as before we have a set of solution pairs (x_1, x_2). The hull of all these solutions provides the hull of Σ^*_{sym}.

1. Solve all 2^4 vertex systems of (22).

2. Solve all possible point systems, where for each of the eight choices of the vertices of $[a_1], [b_1], [b_2]$ we determine a finite number of values taken from $(\underline{a}_2, \overline{a}_2)$, where x_1 and x_2 may plausibly take their maximum or minimum. Up to four such values are generated by (separately) solving the two quadratic equations which are obtained by setting the numerators in (23) and (25) equal to zero, i.e.,

$$b_2 a_2^2 - 2b_1^2 a_2 + a_1 b_1 b_2 = 0, \tag{26}$$

$$b_1 a_2^2 - 2a_1 b_2 a_2 + a_1 b_1^2 = 0. \tag{27}$$

3. Solve all possible point systems, where for each of the eight choices of the vertices of $[a_1], [a_2], [b_2]$ we determine a finite number of values taken from $(\underline{b}_1, \overline{b}_1)$, where x_1 may plausibly take its maximum or minimum. Up to two such values are generated by solving the equation, cf. (24),

$$a_1 b_1^2 - 2a_2^2 b_1 + a_1 a_2 b_2 = 0. \tag{28}$$

4. Solve all possible point systems, where for each of the four choices of the vertices of $[a_1]$ and $[b_2]$ we need to determine a finite number of values taken from $(\underline{a}_2, \overline{a}_2)$ and from $(\underline{b}_1, \overline{b}_1)$, where x_1 may plausibly take its extreme values.

We seek points a_2 and b_1 which jointly satisfy equations (26) and (28). If we solve (28) for b_1 and plug its two solutions into (26), we end up with the condition

$$a_2 c(d - c)(8d + c) = 0,$$

where $c = a_1^2 b_2$ and $d = a_2^3$. Therefore, possibly valid values for a_2 are

$$a_2^{(1)} = 0, \quad a_2^{(2)} = \sqrt[3]{c}, \quad a_2^{(3)} = \frac{1}{2}\sqrt[3]{-c}.$$

However, $c = 0$ is a degenerate case. So if either $a_1 = 0$ or $b_2 = 0$ we must work alternatively:

If $a_1 = 0$, we may conclude from (28) that either $a_2 = 0$ or $b_1 = 0$. However, due to nonsingularity, we have $a_2 \neq 0$. Therefore $b_1 = 0$, and from (26) it follows that $b_2 = 0$, too, whence $0 \in \Sigma^*_{sym}$. Similarly, if $b_2 = 0$, we may conclude from (26) that either $a_2 = 0$ or $b_1 = 0$. If $b_1 = 0$ we have again $0 \in \Sigma^*_{sym}$.

The numbers of point systems to be solved given above are only in the worst case. In general, these will be a lot less. Certainly these numbers are not minimal and can be optimized.

A Method for Global Optimization of Large Systems of Quadratic Constraints

Nitin Lamba, Mark Dietz, Daniel P. Johnson, and Mark S. Boddy

Honeywell Laboratories, Adventium Labs
{nitin.lamba, mark.dietz, daniel.p.johnson}@honeywell.com
mark.boddy@adventiumlabs.org

Abstract. In previous work, we have presented a novel global feasibility solver for the large system of quadratic constraints that arise as subproblems in the solving of hard hybrid problems, such as the scheduling of refineries. In this paper we present the Gradient Optimal Constraint Equation Subdivision (GOCES) algorithm, which incorporates a standard NLP solver and the global feasibility solver to find and establish global optimums for systems of quadratic equations, and present benchmarks.

1 Introduction

We are conducting an ongoing program of research on modeling and solving complex hybrid programming problems (problems involving a mix of discrete and continuous variables), with the end objective of implementing improved hybrid control systems and finite-capacity schedulers for a wide variety of different application domains.

In this report we present an algorithm which is guaranteed either to find the global optimum or to prove global infeasibility for a quadratic system of continuous equations, and show the results of applying the algorithm on standard benchmark problems.

2 Motivation

Prediction and control of physical systems involving complex interactions between a continuous dynamical system and a set of discrete decisions is a common need in a wide variety of application domains. Effective design, simulation and control of such hybrid systems requires the ability to represent and manipulate models including both discrete and continuous components, with some interaction between those components.

For example, constructing a model of refinery operations suitable for scheduling across the whole refinery requires the representation of asynchronous events, time-varying continuous variables, and mode-dependent constraints. In addition, there are important quadratic interrelationships between volume, rate, and time, mass, volume and specific gravity, and among tank volumes, blend volumes and blend qualities.

C. Jermann et al. (Eds.): COCOS 2003, LNCS 3478, pp. 61–70, 2005.
© Springer-Verlag Berlin Heidelberg 2005

This leads to a system containing quadratic constraints with equalities and inequalities.

A refinery planning problem may involve hundreds or thousands of such variables and equations. The corresponding scheduling problem may involve thousands or tens of thousands of variables and constraints. Only recently has the state of the art (and, frankly, the state of the computing hardware) progressed to the point where scheduling the whole refinery on the basis of the individual processing activities themselves has entered the realm of the possible.

One requirement for efficient solution of hybrid problems is the ability to establish global infeasibility of a related set of continuous equations, which allows the algorithms to avoid searching infeasible subsets of the space of possible solutions. Current NLP codes are very efficient at finding local optima for large systems of equations, but suffer from two critical shortcomings in application to non-convex quadratic systems: when they succeed in finding a solution, the solution may only be a local optimum; or more critically for our applications, when they fail to find a solution the problem may in fact have a solution elsewhere in the domain.

In previous years we have developed a method of establishing global infeasibility of large systems of quadratic equations using a combination of enveloping linear programs, bounds propagation methods, and subdivision search: the Gradient Constraint Equation Subdivision (GCES) algorithm [1].

The close correspondence of methods for proving global infeasibility and methods for finding global optimums [2] has subsequently motivated us to extend that method to finding global optimums as well. The result of that effort is the subject of this paper.

3 Subdivision Global Optimality Search

The enhanced version of GCES, called Gradient Optimal Constraint Equation Subdivision (GOCES) has been under development for the past year. The GOCES solver accepts systems of quadratic equations, quadratic inequalities, and continuous variable bounds, and either finds a global optimal solution or establishes global infeasibility of the system, within the normal limits of numerical conditioning, time, and memory. The underlying problem space is NP-Complete, so in general there will be problems for which the algorithm would require exponential time and memory, but the method has proven effective for proving feasibility of large systems of equations (see [1] for further details).

3.1 Overview

The first version of the GCES solver [1] determined global feasibility of a quadratic system of equations, which is polynomial-time equivalent to finding global optimality. Therefore, a logical next step was to extend the solver to add finding global optimality directly to its capabilities. This was achieved in two phases:

1. Replacing the SLP feasibility subroutine (*LLPSS*) by an NLP subroutine capable of finding local optimums within each feasible sub-region.
2. Adding a control structure to the overall subdivision search that is similar in spirit to branch-and-bound.

For the NLP solver used to find a locally optimal solution, we wanted to have some flexibility to try different solvers. To this end, we integrated AMPL[3] with the GOCES solver, so as to have the flexibility of switching various NLP solvers without impacting integration costs due to re-implementing interfaces.

3.2 Basic Algorithm

Let $f(x) : \Re^n \to \Re^m$ be a quadratic function of the form

$$f_k(x) = C_k + \sum_i A_{ki} x_i + \sum_{ij} B_{kji} x_j x_i \ . \tag{1}$$

With an abuse of notation, we shall at times write the function as $f(x) = C + Ax + Bxx$.

We then put upper and lower bounds $lb \in \Re^m, ub \in \Re^m$ on the functions, and lower and upper bounds $u \in \Re^n, v \in \Re^n$ on the variables. (These bounds are allowed to be equal to express equalities.) Without any loss in generality, we can also assume that the first variable x_0 is the objective value. The problem we wish to solve will have the form

$$\min \ x_0 \tag{2}$$
$$\forall k : lb_k \leq f_k(x) \leq ub_k \ .$$
$$\forall i : u_i \leq x_i \leq v_i$$

In the course of solving the problem above, we will be solving a sequence of subsidiary problems. These problems will be parameterized by a trial solution $\bar{x}$ and a set of point bounds $\{u, v\} : u \leq \bar{x} \leq v$.

Given the point bounds, we define the gradient bounds

$$F = (B)_+ u + (B)_- v, \quad G = (B)_+ v + (B)_- u \tag{3}$$

(where the positive and negative parts are taken element-wise over the quadratic tensor) so that whenever $u \leq x \leq v$ we will have $F \leq Bx \leq G$.

The centered representation of a function relative to a given trial solution $\bar{x}$ is

$$f(x) = \overline{C} + \overline{A}(x - \bar{x}) + B(x - \bar{x})(x - \bar{x}) , \tag{4}$$

where $\overline{A} = A + (B + B^*)\bar{x}$, $\overline{C} = C + A\bar{x} + B\bar{x}\bar{x}$. By also defining $\bar{u} = u - \bar{x}$, $\bar{v} = v - \bar{x}$, $\overline{F} = F - B\bar{x}$, and $\overline{G} = G - B\bar{x}$ the bounding inequalities will be equivalent to the centered inequalities

$$\bar{u} \leq x - \bar{x} \leq \bar{v} \tag{5}$$
$$\overline{F} \leq B(x - \bar{x}) \leq \overline{G}.$$

In order to develop our enveloping linear problem, we then bound the quadratic equations on both sides by decomposing $x - \overline{x}$ into two nonnegative variables $z, w \geq 0$, $z - w = x - \overline{x}$, $-w \leq (x - \overline{x})_- \leq 0 \leq (x - \overline{x})_+ \leq z$.

The GCES infeasibility test uses an enveloping linear program known as the Linear Program with Minimal Infeasibility (*LPMI*), which uses one-sided bounds for upper and lower limits on the gradients of the equations within the region. It has the form below (for more details, see [1]).

$$LPMI(\overline{x}, u, v) = \left\{ x : \exists z, w : \begin{array}{l} \min \Sigma(G - F)(z + w) \\ \overline{u} \leq x - \overline{x} \leq \overline{v} \\ lb \leq \overline{C} + \overline{A}(x - \overline{x}) + \overline{G}z - \overline{F}w \\ ub \geq \overline{C} + \overline{A}(x - \overline{x}) + \overline{F}z - \overline{G}w \\ x - \overline{x} = z - w \\ z + w \leq \max(|\overline{v}|, |\overline{u}|) \\ z, w \geq 0 \end{array} \right\} \tag{6}$$

The LPMI rigorously establishes the infeasibility of the original nonlinear constraints.

The enhanced version of GCES currently uses AMPL with CONOPT [3] to determine local optimality/feasibility. As the ranges of variables are subdivided, we have also utilized the continuous constraint propagation methods developed earlier (see [1]) to refine the variable bounds.

The central idea behind the global optimization search is to add an aggressive bound on the objective function value whenever a local optimum is obtained. For instance, in a minimization problem, if Z_{best} is the objective value of the best available solution obtained so far, then the objective upper bound Z_{upper} of all the open nodes in the search tree can be updated as:

$$Z_{upper} = \min(Z_{upper}, Z_{best} - \varepsilon_{opt}) \tag{7}$$

where ε_{opt} is the absolute optimization tolerance set for the system. For a maximization problem, the lower bound of the nodes is updated. As the search proceeds, if all the updated nodes are found to be infeasible using the LPMI, that suffices to prove that there is no better solution than the best available so far (Z_{best}).

The abstract version of the algorithm steps are as follows:

1. Among the current node candidates, choose the node with initial trial solution which has the minimal *max infeasibility*. If there are no open nodes left in the search tree, then success. Return with the best available solution.
2. Use bound propagation through the constraints to find refined bounds. If the resulting bounds are infeasible, declare the node infeasible.
3. Evaluate *LPMI* using CPLEX for infeasibility; if so, declare the node infeasible.
4. Use AMPL/CONOPT to find a feasible solution within the current node. If that solution is a local optimum, update the best solution and the objective bounds for all the open nodes and re-propagate them.

5. Update the trial solution with the solution obtained from CONOPT solver and evaluate *LPMI* again for infeasibility; if so, declare the node infeasible.
6. If (4) is successful (found a locally optimum solution), split the point optimal node into multiple sub-nodes by subdividing the range of the objective variable. Project the trial solution into each region, and go to step (1).
7. If (2), (3), (5) are inconclusive for the node, split the point optimal node into multiple sub-nodes by subdividing the range of a chosen variable. Project the trial solution into each region, and go to (1).

In later sections, we detail the strategies to choose a node in step (1) and a variable in step (7).

3.3 Search Strategies

The solver makes two main decisions on how to subdivide a node: which variable to divide and how to divide the range for that variable. In developing the feasibility solver GCES we investigated eight strategies for choosing the variable to split, reported in [1]. In developing global optimality, we restricted our attention to three strategies, covering the range from best worst-case behavior to most adaptive, which we have found to be most effective.

The first strategy, Strategy L, uses the trial solution retrieved from CONOPT and chooses the constraint with maximum infeasibility and then chooses the variable with the largest bounds. The second strategy, Strategy B, uses the trial solution retrieved from the final LPMI solved by CPLEX to choose the quadratic constraint k with the highest infeasibility. This strategy chooses a variable x_i, in the constraint which maximizes $(G_{ki} - F_{ki}) * (v_i - u_i)$. Then the variable x_j with the largest coefficient b_{ij} in constraint k is chosen as the final variable to split. The third strategy, Strategy K, uses the trial solution retrieved from the final LPMI solved by CPLEX to choose the quadratic constraint k with the highest infeasibility. The strategy chooses the quadratic variable x_i that maximizes $z_i + w_i$.

Once a variable has been chosen, its range must be divided in some way. Tests were run using two strategies for range splitting. Both range splitting strategies used Strategy K for variable selection in these test runs.

The first strategy, R2, divides the variable range into three regions:

$$u \leq x \leq \bar{x} - \delta \qquad (8)$$
$$\bar{x} - \delta \leq x \leq \bar{x} + \delta$$
$$\bar{x} + \delta \leq x \leq v$$

where u is lower bound, v is upper bound, $\bar{x}$ is the trial solution, and δ is computed to give the desired size of the center region.

If the trial solution is closer than epsilon to the lower bound, then the range is divided into two regions, $u \leq x \leq \bar{x} + \delta$ and $\bar{x} + \delta \leq x \leq v$. Similarly if the trial solution is near the upper bound, the regions are $u \leq x \leq \bar{x} - \delta$ and $\bar{x} - \delta \leq x \leq v$.

The second strategy, R4, divides the variable range into n regions of equal size. It also splits the subdivision containing $\bar{x}$ into two divisions around $\bar{x}$. These divisions overlap by 10ε in order to insure round off error does not rule out a solution.

In our development process, we typically use R4 to test variable choice strategies, because the resulting splits are only dependent on the variable bounds, and not on the numerical values of the particular local solutions. The resulting search patterns are much less sensitive to numerical vagaries of the CPLEX and NLP codes.

3.4 Benchmarks

We used a series of polynomial benchmarks taken from Floudas and Pardalos [4]. The equations were rewritten as linear combinations of quadratic terms by adding variables where necessary. Table 1 below summarizes some statistics of the problems.

Three variable selection strategies were run using the equal split strategy R4, n = 5. Table 2 summarizes the results. Strategy L is a simple, but ineffective strategy. For test problem F2.8, a fairly easy problem for the other strategies, an incorrect solution was reached due to the extra computation. Strategy B and Strategy K performed similarly on most of the problems with an overall all edge to Strategy K.

Table 3 below summarizes the results with R2, and n = 5 for R4, using variable choice strategy K.

Table 1. Benchmark Problems

m = number of equations
n = number of variables
nz = non-zero entries in Jacobian
mq = number of equations with quadratic terms
nq = number of variables appearing in quadratic terms

Problem Number	Type	m	n	nz	mq	nq
F2.1	Quadratic programming	10	14	30	5	5
F2.1	Quadratic programming	10	14	30	5	5
F2.2	Quadratic programming	11	15	34	5	5
F2.3	Quadratic programming	17	21	57	4	4
F2.4	Quadratic programming	10	11	44	1	1
F2.5	Quadratic programming	22	21	136	7	7
F2.8	Quadratic programming	38	52	149	24	24
F3.1	Quadratically constrained	15	17	40	5	8
F3.2	Quadratically constrained	18	17	49	8	5
F3.3	Quadratically constrained	16	16	43	6	6
F5.4	Blending/Pooling/Separation	72	78	156	18	18
F6.2	Pooling/Blending	12	15	37	2	3
F6.3	Pooling/Blending	12	15	37	2	3
F6.4	Pooling/Blending	12	15	37	2	3

Strategy R2 tends to perform better than R4. R2 is able to rule out larger regions where R4 would divide the region into more parts and then have to rule out each individually.

Table 2. Number of Nodes Generated For Variable Choice Strategies

Test Problem	Problem type	Number of Nodes Generated Strategy L	Strategy B	Strategy K
F2.1	Quadratic programming	232	58	58
F2.2	Quadratic programming	1	1	1
F2.3	Quadratic programming	9	9	9
F2.4	Quadratic programming	121	14	14
F2.5	Quadratic programming	9	9	9
F2.8	Quadratic programming	24650*	78	78
F3.1	Quadratically constrained	63089	7685	3051
F3.2	Quadratically constrained	9	9	9
F3.3	Quadratically constrained	133*	60	60
F5.4	Blending/Pooling/Separation	1084780	24617	22943
F6.2	Pooling/Blending	2838	444	274
F6.3	Pooling/Blending	8212	576	620
F6.4	Pooling/Blending	765	96	137

* indicates suboptimal or infeasible solution given

Table 3. Number of Nodes Generated For Range Splitting Strategies

Test Problem	Problem type	Number of Nodes Strategy R2	Strategy R4
F2.1	Quadratic programming	24	58
F2.2	Quadratic programming	1	1
F2.3	Quadratic programming	3	9
F2.4	Quadratic programming	10	14
F2.5	Quadratic programming	3	9
F2.8	Quadratic programming	61	78
F3.1	Quadratically constrained	790	3051
F3.2	Quadratically constrained	3	9
F3.3	Quadratically constrained	16	60
F5.4	Blending/Pooling/Separation	9270	22943
F6.2	Pooling/Blending	91	274
F6.3	Pooling/Blending	158	620
F6.4	Pooling/Blending	68	137

3.5 Optimality Tolerance

GOCES adds an aggressive bound on the objective function value whenever a local optimum is obtained. Whenever a local optimum is found, GOCES then imposes a new constraint on the objective value, $\min Z : Z \leq Z_{loc} - \varepsilon_{opt}$, and seeks to either prove that the resulting system is infeasible (e.g. Z_{loc} is the global optimal objective value), or finds a objective value better than the old local optimum by at least the optimality tolerance. Hence the algorithm returns an objective value Z_{fin} and a feasible point X_{fin} at which that objective is achieved, and a guarantee that the true global optimal objective is no better than $Z_{fin} - \varepsilon_{opt}$.

However, as the optimality tolerance ε_{opt} is reduced, it becomes more difficult to prove infeasibility. This effect appears to be due to the clustering problem [5], where finer and finer subdivisions are necessary in the vicinity of the current candidate for global optimum. Table 4 lists the variation in the total number of nodes with ε_{opt} while solving the test problems. Our testing shows a "critical threshold" effect, where the increase in computation time as the tolerance decreases is not particularly troublesome, until a threshold is reached, and the time necessary increases past the point at which we terminated our runs (several hours of runtime).

To understand the global optimum tolerance, consider test problem F2.1, which had multiple local optima. With $\varepsilon_{opt} = 10^{-1}$, it found an "optimal" value at -16.5 and showed that there was no solution with a value better than $-16.5 * (1 + \varepsilon_{opt}) = 18.15$. The global optimum was at -17.0, for an improvement of 3%. When the optimality tolerance was reduced to 10^{-2}, the GOCES found the true global optimum.

Table 4. Number of Nodes Generated For Different Optimality Tolerances

Test Problem	Problem type	Number of Nodes				
		tol = 10^{-5}	tol = 10^{-4}	tol = 10^{-3}	tol = 10^{-2}	tol = 10^{-1}
F2.1	Quadratic programming	58	58	58	58	58
F2.2	Quadratic programming	9	9	9	1	1
F2.3	Quadratic programming	9	9	9	9	9
F2.4	Quadratic programming	14	14	14	14	14
F2.5	Quadratic programming	27	27	27	15	9
F2.8	Quadratic programming		1347	1002	409	78
F3.1	Quadratically constrained	17488	16842	12510	7894	3051
F3.2	Quadratically constrained	51	51	51	39	9
F3.3	Quadratically constrained	74	74	74	68	60
F5.4	Blending/Pooling/Separation	21129	25538	21820	22943	
F6.2	Pooling/Blending		339	314	284	274
F6.3	Pooling/Blending	1051	622	674	601	620
F6.4	Pooling/Blending		143	143	143	137

3.6 Other Practical Aspects

The solver is coded in Java with an interface to the CPLEX linear programming library and an interface to AMPL for non-linear local optimizations. It takes as input the expected magnitudes of the variables and objective function, which are used for scaling.

3.7 Benchmark Timing

As discussed in Section 3.4, we used a series of polynomial benchmarks taken from Floudas and Pardalos [4]. At the present time we lack access to timings for other comparable algorithms, so we present our timings only. Table 5 below summarizes some statistics of the problems for strategy K-R2, tol = 10^{-4}.

Execution time: Runs were made on a x86 class desktop PC (1.99 GHz Pentium 4, 512 Mb RAM) running Microsoft Windows XP Professional.

Local Optima Searched: The number of local (including the global) optima found before global optimality was established.

Nodes Searched: Number of subdomains examined.

LPMI Problems Executed: Number of enveloping linear programs executed.

CONOPT Minor Iterations: Number of times CONOPT iterated (e.g. found a new search point and updated its gradient estimates).

Table 5. Benchmark Results

Test Problem	Execution Time (ms)	Local Optima Searched	Nodes Searched	LPMI Problems Executed	CONOPT Minor Iterations
F2.1	3626	2	29	39	91
F2.2	1012	1	3	3	11
F2.3	1121	1	3	3	4
F2.4	1643	1	10	11	21
F2.5	2915	1	25	29	58
F2.8	21001	4	201	278	1028
F3.1	329105	1	4854	5711	8807
F3.2	4276	1	44	57	63
F3.3	4997	4	24	36	71
F5.4	1058133	1	13132	12531	37614
F6.2	13910	2	160	181	255
F6.3	13269	2	159	160	281
F6.4	7871	3	77	73	151

The new version of the solver was also tested on the scheduling problem of refinery operations discussed in [1]. The problem consists of 6,771 variables with 8,647 constraints and equations, with 1,976 of the equations being quadratic, reflecting chemical distillation and blending equations (problem 5.4 above is the largest benchmark we tested, consisting of 48 constraints, 54 variables, and 18 quadratic equations). The solver found an optimum of 0.44 and proved that there was no solution with value greater than 0.84 but could not improve on that overnight. Difficulties identified were the scale of the problem and solution clustering near the optimum solution.

4 Future Work

Nonlinear Functions: Given the fact that we have a solver that can work with quadratic constraints, the current implementation can handle an arbitrary polynomial or rational function through rewriting and the introduction of additional variables. The issue is a heuristic one (system performance), not an expressive one. The GOCES framework can be extended to include any nonlinear function for which one has analytic gradients, and for which one can compute reasonable function and gradient bounds given variable bounds.

Efficiency: While we are constantly improving the performance of the search through various pragmatic measures, there is much yet to be done. In addition to further effort in the areas listed here, we intend to investigate the use of more sophisticated scaling techniques, and effective utilization of more problem information that is generally available in nonlinear solvers.

Other Application Areas: The current hybrid solver is intended to solve scheduling problems. Other potential domains that we wish to investigate include batch manufacturing, satellite and spacecraft operations, transportation and logistics planning, abstract planning problems, and the control of hybrid systems, and linear hybrid automaton (LHA).

5 Summary

We have extended our global equation solver to a global optimizer for system of quadratic constraints capable of modeling and solving scheduling problems involving an entire petroleum refinery, from crude oil deliveries, through several stages of processing of intermediate material, to shipments of finished product. This scheduler employs the architecture described previously [6] for the coordinated operation of discrete and continuous solvers. There is considerable work remaining on all fronts, especially improvement in the search algorithm.

References

[1] Boddy, M., and Johnson, D.: A New Method for the Global Solution of Large Systems of Continuous Constraints, COCOS Proceedings, 2002.
[2] Papadimitriou, C., and Steiglitz, K.: Combinatorial Optimization Algorithms and Complexity, Dover Publications, 1998
[3] CONOPT solver by ARKI Consulting & Development A/S (http://www.conopt.com)
[4] Floudas, C., and Pardalos, P.: A Collection of Test Problems for Constrained Global Optimization Algorithms, Lecture Notes in Computer Science # 455, Springer-Verlag, 1990
[5] Bliek, C. *et al.*,: COCONUT Deliverable D1 - Algorithms for Solving Nonlinear and Constrained Optimization Problems, The COCONUT Project, 2001 (http://www.mat.univie.ac.at/~neum/glopt/coconut/)
[6] Boddy, M., and Krebsbach, K.: Hybrid Reasoning for Complex Systems, 1997 Fall Symposium on Model-directed Autonomous Systems.

A Comparison of Methods for the Computation of Affine Lower Bound Functions for Polynomials*

Jürgen Garloff and Andrew P. Smith

University of Applied Sciences / FH Konstanz,
Postfach 100543, D-78405 Konstanz, Germany
{garloff, smith}@fh-konstanz.de

Dedicated to Professor Dr. Karl Nickel on the occasion of his eightieth birthday.

Abstract. In this paper the problem of finding an affine lower bound function for a multivariate polynomial is considered. For this task, a number of methods are presented, all based on the expansion of the given polynomial into Bernstein polynomials. Error bounds and numerical results for a series of randomly-generated polynomials are given.

Keywords: Bernstein polynomials, control points, convex hull, bound functions, complexity, global optimization.

1 Introduction

Finding a convex lower bound function for a given function is of paramount importance in global optimization when a branch and bound approach is used. Of special interest are convex envelopes, i.e., uniformly best underestimating convex functions, cf. [5], [15], [21].

Because of their simplicity and ease of computation, constant and affine lower bound functions are especially useful. Constant bound functions are thoroughly used when interval computation techniques are applied to global optimization, cf. [10], [13], [20]. However, when using constant bound functions, all information about the shape of the given function is lost. A compromise between convex envelopes, which require in the general case much computational effort, and constant lower bound functions are affine lower bound functions.

Here we concentrate on such bound functions for multivariate polynomials. These bound functions are constructed from the coefficients of the expansion of the given polynomial into Bernstein polynomials. Properties of Bernstein polynomials are introduced in Section 2; the reader is also referred to [4], [6], [18], [22]. In Section 3 we present a number of variant methods, together with a suitable transformation that may be applied to improve the results. Numerical results for a series of randomly-generated polynomials are given in Section 4, with a comparison of the error bounds.

---

* This work has been supported by the German Research Council (DFG).

C. Jermann et al. (Eds.): COCOS 2003, LNCS 3478, pp. 71–85, 2005.

2   Bernstein Polynomials and Notation

We define multiindices $i = (i_1, \ldots, i_n)^T$ as vectors, where the n components are nonnegative integers. The vector 0 denotes the multiindex with all components equal to 0, which should not cause ambiguity. Comparisons are used entrywise. Also the arithmetic operators on multiindices are defined componentwise such that $i \odot l := (i_1 \odot l_1, \ldots, i_n \odot l_n)^T$, for $\odot = +, -, \times$, and $/$ (with $l > 0$). For instance, i/l, $0 \le i \le l$, defines the Greville abscissae. For $x \in \mathbf{R}^n$ its multipowers are

$$x^i := \prod_{\mu=1}^{n} x_\mu^{i_\mu}. \tag{1}$$

Multipowers of multiindices are not required here; instead we shall write $i^0, \ldots, i^n$ for a sequence of $n+1$ multiindices. For the sum we use the notation

$$\sum_{i=0}^{l} := \sum_{i_1=0}^{l_1} \cdots \sum_{i_n=0}^{l_n}. \tag{2}$$

A multivariate polynomial p of degree $l = (l_1, \ldots, l_n)^T$ can be represented as

$$p(x) = \sum_{i=0}^{l} a_i x^i \quad \text{with} \quad a_i \in \mathbf{R}, 0 \le i \le l, \text{ and } a_l \ne 0. \tag{3}$$

The ith Bernstein polynomial of degree l is

$$B_i(x) := \binom{l}{i} x^i (1-x)^{l-i}, \tag{4}$$

where the generalized binomial coefficient is defined by $\binom{l}{i} := \prod_{\mu=1}^{n} \binom{l_\mu}{i_\mu}$, and x is contained in the unit box[1] $I = [0,1]^n$. It is well-known that the Bernstein polynomials form a basis in the space of multivariate polynomials, and each polynomial in the form (3) can be represented in its Bernstein form over I

$$p(x) = \sum_{i=0}^{l} b_i B_i(x), \tag{5}$$

where the *Bernstein coefficients* b_i are given by

$$b_i = \sum_{j=0}^{i} \frac{\binom{i}{j}}{\binom{l}{j}} a_j \quad \text{for} \quad 0 \le i \le l. \tag{6}$$

---

[1] Without loss of generality we consider in the sequel the unit box since any nonempty box in $\mathbf{R}^n$ can be mapped affinely thereupon. For the respective formulae for general boxes in the univariate case see [19] and their extensions to the multivariate case, e.g., Section 7.3.2 in [2].

A fundamental property for our approach is the *convex hull property*

$$\left\{ \begin{pmatrix} x \\ p(x) \end{pmatrix} : x \in I \right\} \subseteq conv \left\{ \begin{pmatrix} i/l \\ b_i \end{pmatrix} : 0 \leq i \leq l \right\}, \tag{7}$$

where the convex hull is denoted by *conv*. The points $\begin{pmatrix} i/l \\ b_i \end{pmatrix}$ are called *control points* of p. The enclosure (7) yields the inequalities

$$\min\{b_i : 0 \leq i \leq l\} \leq p(x) \leq \max\{b_i : 0 \leq i \leq l\} \tag{8}$$

for all $x \in I$. For ease of presentation we shall sometimes simply use b_i to denote the control point associated with the Bernstein coefficient b_i, where the context should make this unambiguous. Exponentiation on control points, Bernstein coefficients, or vectors is also not required here; therefore $b^0, \ldots, b^n$ is a sequence of $n + 1$ control points or Bernstein coefficients (with $b^j = b_{i^j}$), and $u^1, \ldots, u^n$ is a sequence of n vectors.

3   Affine Lower Bound Functions

In this section we explore a number of different methods for the computation of affine lower bound functions for polynomials. In each case it is assumed that we have a multivariate polynomial p given by (3) and that its Bernstein coefficients b_i, $0 \leq i \leq l$, have been computed.

Theorems 1 and 2 below are independent of any particular method. They characterize an affine lower bound function as the solution of a linear programming problem. There is a degree of freedom in that the statements contain an index set $\hat{J}$ which corresponds to a facet of the convex hull of the control points of p. According to the choice of $\hat{J}$ and the way in which the linear programming problem is posed (either all inequalities in (12) are considered or only a few), numerous related methods can be designed. We discuss a few in the sequel.

3.1   Method 1

Constant bound functions can be computed easily and cheaply from the Bernstein coefficients: The left-hand side of (8) implies that the constant function provided by the minimum Bernstein coefficient

$$c_0(x) = b_{i^0} = \min\{b_i : 0 \leq i \leq l\} \tag{9}$$

is an affine lower bound function for the polynomial p given by (3) over the unit box I. However, due to the lack of shape information, these bound functions usually perform relatively poorly.

3.2   Method 2

This method was presented in [7] and relies on the following construction: Choose a control point b_{i^0} with minimum Bernstein coefficient, cf. (9). Let $\hat{J}$ be a set of at least n multiindices such that the slopes between b_{i^0} and the control points with Greville abscissae associated with $\hat{J}$ are smaller than or equal to the slopes between b_{i^0} and the remaining control points. Then the desired affine lower bound function is provided as the solution of the linear programming problem to maximize the affine function at the Greville abscissae associated with $\hat{J}$ under the constraints that this affine function remains below all control points and passes through b_{i^0}. More precisely, the following theorem holds true.

Theorem 1. *Let $\{b_i\}_{i=0}^l$ denote the Bernstein coefficients of the polynomial p given by (3). Choose i^0 as in (9) and let $\hat{J} \subseteq \{\hat{j} : 0 \leq \hat{j} \leq l, \hat{j} \neq i^0\}$ be a set of at least n multiindices such that*

$$\frac{b_{\hat{j}} - b_{i^0}}{\|\hat{j}/l - i^0/l\|} \leqslant \frac{b_i - b_{i^0}}{\|i/l - i^0/l\|} \text{ for each } \hat{j} \in \hat{J}, \ 0 \leq i \leq l, \ i \neq i^0, \ i \notin \hat{J}. \quad (10)$$

Here, $\|\cdot\|$ denotes some vector norm. Then the linear programming problem

$$\min \ (\sum_{\hat{j} \in \hat{J}} (\hat{j}/l - i^0/l))^T \cdot s \qquad \text{subject to} \quad (11)$$

$$(i/l - i^0/l)^T \cdot s \geq b_{i^0} - b_i \text{ for } 0 \leq i \leq l, i \neq i^0 \quad (12)$$

has the following properties:

1. *It has an optimal solution $\hat{s}$.*
2. *The affine function*

$$c(x) := -\hat{s}^T \cdot x + (\hat{s}^T \cdot (i^0/l) + b_{i^0}) \quad (13)$$

is a lower bound function for p on I.

In the univariate case, by definition (10), $\hat{J}$ can be chosen such that it consists of exactly one element $\hat{j}$ which may not be uniquely defined. The slope of the affine lower bound function c is equal to the smallest possible slope between the control points. Moreover, the optimal solution of the linear programming problem (11) and (12) can be given explicitly in the univariate case.

Theorem 2. *Suppose that all assumptions of Theorem 1 are satisfied, where $n = 1$ and where $\|\cdot\|$ denotes the absolute value. Choose $\hat{J} = \{\hat{j}\}$, where $\hat{j}$ satisfies*

$$\frac{b_{\hat{j}} - b_{i^0}}{|\hat{j}/l - i^0/l|} = \min \left\{ \frac{b_i - b_{i^0}}{|i/l - i^0/l|} : 0 \leq i \leq l, \ i \neq i^0 \right\}.$$

There then exists an optimal solution $\hat{s}$ of the linear programming problem (11), (12) which satisfies

$$\hat{s} = -\frac{b_{\hat{j}} - b_{i^0}}{\hat{j}/l - i^0/l}. \quad (14)$$

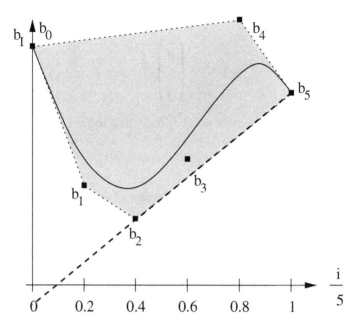

Fig. 1. The curve of a polynomial of fifth degree (bold), the convex hull (shaded) of its control points (marked by squares), and an affine lower bound function constructed as in Theorem 2

Figure 1 illustrates the construction of such an affine lower bound function.

In the univariate case the computational work for constructing such bound functions is negligible, but in the multivariate case a linear programming problem has to be solved. In the branch and bound framework it may happen that one has to solve subproblems on numerous subboxes of the starting region, so that for higher dimensions solving the linear programming problems becomes a computational burden.

3.3   Method 3

Overview. This method was introduced in [9]. It only requires the solution of a system of linear equations together with a sequence of back substitutions. The following construction aims to find hyperplanes passing through the control point b^0 (associated with the minimum Bernstein coefficient b_{i^0}, cf. (9)) which approximate from below the lower part of the convex hull of the control points increasingly well. In addition to b^0, we designate n additional control points $b^1, \ldots, b^n$. Starting with c_0, cf. (9), we construct from these control points a sequence of affine lower bound functions $c_1, \ldots, c_n$. We end up with c_n, a hyperplane which passes through a lower facet of the convex hull spanned by the control points $b^0, \ldots, b^n$. In the course of this construction, we generate a set of linearly independent vectors $\{u^1, \ldots, u^n\}$ and we compute slopes from b^0 to b^j in direction u^j. Also, w^j denotes the vector connecting b^0 and b^j.

Algorithm - First Iteration:

$$Let \ u^1 = \begin{pmatrix} 1 \\ 0 \\ \vdots \\ 0 \end{pmatrix}.$$

Compute slopes g_i^1 from the control point b_i to b^0 in direction u^1:

$$g_i^1 = \frac{b_i - b^0}{\frac{i_1}{l_1} - \frac{i_1^0}{l_1}} \quad \text{for all } i \text{ with } i_1 \neq i_1^0.$$

Let i^1 be a multiindex with smallest absolute value of associated slope g_i^1. Designate the control point $b^1 = \left(\frac{i^1}{l}, b_{i^1}\right)^T$, the slope $\alpha_1 = g_{i^1}^1$, and the vector $w^1 = \frac{i^1 - i^0}{l}$. Define the lower bound function

$$c_1(x) = b^0 + \alpha_1 u^1 \cdot \left(x - \frac{i^0}{l}\right).$$

Algorithm - jth Iteration, $j = 2, \ldots, n$:

$$Let \ \tilde{u}^j = \begin{pmatrix} \beta_1^j \\ \vdots \\ \beta_{j-1}^j \\ 1 \\ 0 \\ \vdots \\ 0 \end{pmatrix}$$

$$\text{such that } \tilde{u}^j \cdot w^k = 0, \ k = 1, \ldots, j - 1. \tag{15}$$

Normalize this vector thusly:

$$u^j = \frac{\tilde{u}^j}{\|\tilde{u}^j\|}. \tag{16}$$

Compute slopes g_i^j from the control point b_i to b^0 in direction u^j:

$$g_i^j = \frac{b_i - c_{j-1}\left(\frac{i}{l}\right)}{\frac{i - i^0}{l} \cdot u^j} \quad \text{for all } i, \text{ except where } \frac{i - i^0}{l} \cdot u^j = 0. \tag{17}$$

Let i^j be a multiindex with smallest absolute value of associated slope g_i^j. Designate the control point $b^j = \left(\frac{i^j}{l}, b_{i^j}\right)^T$, the slope $\alpha_j = g_{i^j}^j$, and the vector $w^j = \frac{i^j - i^0}{l}$. Define the lower bound function

$$c_j(x) = c_{j-1}(x) + \alpha_j u^j \cdot \left(x - \frac{i^0}{l}\right). \tag{18}$$

Remark: Solving (15) for the coefficients $\beta_1^j, \ldots, \beta_{j-1}^j$ requires the solution of a system of $j-1$ linear equations in $j-1$ unknowns. This system has a unique solution due to the linear independence amongst the vectors $w^1, \ldots, w^n$, as proven in [9].

For the n iterations of the above algorithm, the solution of such a sequence of systems of linear equations would normally require $\frac{1}{6}n^4 + O(n^3)$ arithmetic operations. However we can take advantage of the fact that, in the jth iteration, the vectors $w^1, \ldots, w^{j-1}$ are unchanged from the previous iteration. The solution of these systems can then be formulated as Gaussian elimination applied rowwise to the single $(n-1) \times (n-1)$ matrix whose rows consist of the vectors $w^{n-1,1}, \ldots, w^{n-1,n-1}$ and right-hand side $-(w_n^1, \ldots, w_n^{n-1})^T$. In addition, a sequence of back-substitution steps has to be performed. Then altogether only $n^3 + O(n^2)$ arithmetic operations are required.

Let

$$L = \sqrt[n]{\prod_{i=1}^{n} (l_i + 1)}.$$

There are then L^n Bernstein coefficients, so that the computation of the slopes g_i^j (17) in all iterations requires at most $n^2 L^n + L^n O(n)$ arithmetic operations. This new approach therefore requires less computational effort in general than Method 2, which is based on the solution of a linear programming problem with upto $L^n - 1$ constraints. [2]

The following results were given in [9]:

Theorem 3. *With the notation of the above algorithm, it holds for all* $j = 0, \ldots, n$ *that*

$$c_j \left(\frac{i^k}{l} \right) = b^k, \quad \text{for } k = 0, \ldots, j.$$

In particular, we have that

$$c_n \left(\frac{i^k}{l} \right) = b^k, \quad k = 0, \ldots, n, \tag{19}$$

which means that c_n passes through all $n+1$ control points $b^0, \ldots, b^n$. Since c_n is by construction a lower bound function, $b^0, \ldots, b^n$ must therefore span a lower facet of the convex hull of all control points.

We obtain a pointwise error bound for the underestimating function c_n which also holds true for c_n replaced by the affine lower bound function c constructed by Method 2, cf. [7].

Theorem 4. *Let* $\{b_i\}_{i=0}^{l}$ *denote the Bernstein coefficients of the polynomial* p *given by (3). Then the affine lower bound function* c_n *satisfies the a posteriori error bound*

$$0 \leq p(x) - c_n(x) \leq \max \left\{ b_i - c_n \left(\frac{i}{l} \right) : 0 \leq i \leq l \right\}, \quad x \in I. \tag{20}$$

---

[2] In our computations, we have chosen exactly $L^n - 1$ constraints.

In the univariate case, this error bound specifies to the following bound which exhibits quadratic convergence with respect to the width of the intervals, see [7].

Theorem 5. *Suppose $n = 1$ and that the assumptions of Theorem 4 hold, then the affine lower bound function c_n satisfies the error bound ($x \in I$)*

$$0 \le p(x) - c_n(x) \le \max \left\{ \left(\frac{b_i - b^0}{\frac{i}{l} - \frac{i^0}{l}} - \frac{b^1 - b^0}{\frac{i^1}{l} - \frac{i^0}{l}} \right) \left(\frac{i}{l} - \frac{i^0}{l} \right) : 0 \le i \le l, \ i \ne i^0 \right\}.$$

Theorem 5 also holds true for c_n replaced by the affine lower bound function c of Method 2 and i^1 replaced by $\hat{j}$, cf. Theorem 2.

It was shown in [7] and [9] that affine polynomials coincide with their affine lower bound functions constructed therein. This suggests that almost affine polynomials should be approximated rather well by their affine lower bound functions. This is confirmed by our numerical experiences.

In [8] we introduced a lower bound function for univariate polynomials which is composed of two affine lower bound functions. The extension to the multivariate case is as follows: In each step, compute slopes as before, but select α_j^- as the greatest negative g_i^j value, and α_j^+ as the smallest positive g_i^j value. From each previous lower bound function c_{j-1}, generate two new lower bound functions, using α_j^- and α_j^+. Instead of a sequence of functions, we now obtain after n iterations upto 2^n lower bound functions due to the binary tree structure.

It is worth noting that in the current version of our algorithm the choice of the direction vectors u^j (16) is rather arbitrary. However our numerical experience suggests that this may influence the resultant bound function (i.e. which lower facet of the convex hull of the control points is emulated). A future modification to the algorithm may therefore use a simple heuristic function to choose these vectors in an alternative direction such that a more suitable facet of the lower convex hull is designated. With the orthogonality requirement (15), there are $n - j$ degrees of freedom in this selection.

3.4   Methods 4 and 5

We also propose two simpler methods for the construction of affine lower bound functions based on the Bernstein expansion, with the computation of slopes and differences only, with still lower complexity. Method 4 is based on a choice of control points corresponding to $n + 1$ smallest Bernstein coefficients and Method 5 is based on a choice of a control point corresponding to the minimum Bernstein coefficient and n others which connect to it with minimum absolute value of gradient. In both cases, a lower bound function interpolating the designated control points is computed, requiring the solution of a single system of linear equations. A degenerate case may arise when this system has no unique solution — with the terminology of Method 3, the set of vectors $\{w_j\}$ is linearly dependent. Such cases are tested for and excluded from consideration during the designation of the control points.

Additionally, both methods (unmodified) are not guaranteed to deliver a valid lower bound function — exceptionally there may still occur control points below it. Therefore an error term (20) is computed. If this is negative, it is necessary to adjust the bound function by a downward shift: the absolute value of this error is subtracted from its constant term.

As will become evident from the numerical results in the following section, both of these methods may perform unexpectedly poorly under certain configurations of control points. Two such examples are illustrated in the following figures, where the small circles are the control points of a bivariate polynomial. Those control points filled in black are those which are designated, leading to the construction of a lower bound function (the shaded plane), in the first case after a necessary downward shift. Although both methods usually deliver a bound function with correct shape information (i.e. an improvement over Method 1), this is seen not always to be the case. For this reason, there are no worthwhile error bounds that can be presented for these two methods.

3.5   An Equilibriation Transformation

A limitation of all the above methods is that the resultant lower bound function must pass through the minimum control point b_{i^0} (except in cases where a downward shift is necessary for Methods 4 and 5). Whilst this is often a good choice, it is not always so. Figure 4 gives a simple example where the optimal lower bound function does not in fact pass through the minimum control point. In this case it would seem sensible to utilise the shape information provided by a broad spread of the control points (global shape information over the box) in addition to that already given by a small number of specially designated control points (which may be clustered) as per the above algorithms (local shape information near the minimum control point). We can lift the restriction that the lower bound function must pass through b_{i^0}. Indeed, if there are many Bernstein coefficients (i.e. for polynomials of high degree) the global shape information may be at least as important, if not more so, as the local information. This is especially evident in the cases where Methods 4 and 5 perform poorly (see Figures 2 and 3).

To this end, we can envisage the determination of the lower bound function as a three-stage process. Firstly, we apply an affine transformation to the control points, which we call the *equilibriation transformation*, derived from the control points on the edges of the box, and approximating the global shape information. Secondly, we compute an affine lower bound function c^* for the transformed polynomial p^* (and its control points b_i^*), by using one of Methods 1-5 above. Lastly, we apply the transformation in reverse to obtain an affine lower bound function c for the original polynomial.

We define the equilibriation transformation on the control points as follows:

$$b_i \; \mapsto \; b_i^* := b_i - \sum_{j=1}^{n} \frac{i_j}{l_j} \left(b_{\left(\lfloor \frac{l_1}{2} \rfloor, \ldots, l_j, \ldots, \lfloor \frac{l_n}{2} \rfloor \right)} - b_{\left(\lfloor \frac{l_1}{2} \rfloor, \ldots, 0, \ldots, \lfloor \frac{l_n}{2} \rfloor \right)} \right), \quad 0 \le i \le l.$$

After applying this transformation, the global shape (i.e. the shape over the whole box) of the polynomial has been approximately flattened, i.e.

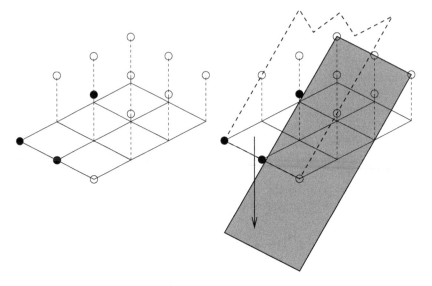

Fig. 2. Method 4 - Example of poor lower bound function

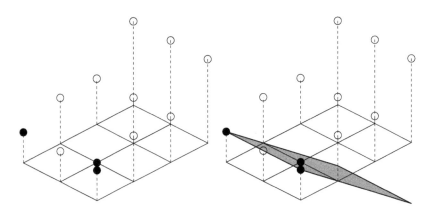

Fig. 3. Method 5 - Example of poor lower bound function

$$b^*_{\left(0,\lfloor \frac{l_2}{2}\rfloor,\ldots,\lfloor \frac{l_n}{2}\rfloor\right)} = b^*_{\left(l_1,\lfloor \frac{l_2}{2}\rfloor,\ldots,\lfloor \frac{l_n}{2}\rfloor\right)},$$

$$\vdots$$

$$b^*_{\left(\lfloor \frac{l_1}{2}\rfloor,\ldots,0,\ldots,\lfloor \frac{l_n}{2}\rfloor\right)} = b^*_{\left(\lfloor \frac{l_1}{2}\rfloor,\ldots,l_j,\ldots,\lfloor \frac{l_n}{2}\rfloor\right)},$$

$$\vdots$$

$$b^*_{\left(\lfloor \frac{l_1}{2}\rfloor,\ldots,\lfloor \frac{l_n-1}{2}\rfloor,0\right)} = b^*_{\left(\lfloor \frac{l_1}{2}\rfloor,\ldots,\lfloor \frac{l_n-1}{2}\rfloor,l_n\right)}.$$

The effect of this transformation is illustrated in Figure 4 with a univariate polynomial of degree 6, yielding an optimal bound function which does not pass through the minimum control point.

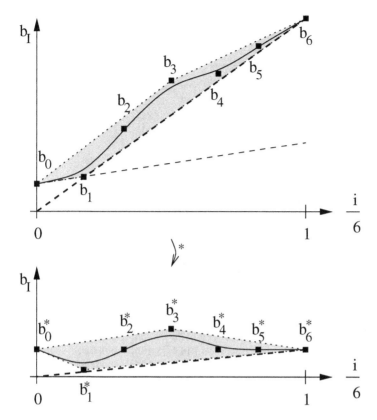

Fig. 4. Result of applying the equilibriation transformation; the improved/transformed bound function is given in bold dashed

3.6   Verification

Due to rounding errors, inaccuracies may be introduced into the calculation of the Bernstein coefficients and the lower bound functions. Especially it may happen that the computed lower bound function value is greater than the corresponding original function value. This may lead to erroneous results in applications. Suggestions for the way in which one can obtain functions which are guaranteed to be lower bound functions also in the presence of rounding errors are given in [7]. One such approach is to compute an error term (20) followed by a downward shift, if necessary, as in Methods 4 and 5. For a different approach see [3], [11], [14].

4   Examples

The above methods for computing lower bound functions, both with and without the equilibriation transformation, were tested with a number of multivariate

Table 1. Results for random polynomials

	Method			1 (Constant/Affine)					
n	D	k	$(D+1)^n$	time (s)	δ	δ_E	time (s)	δ	δ_E
2	2	5	9	0.000040	1.414	0.777			
2	6	10	49	0.00013	1.989	1.570			
2	10	20	121	0.00039	2.867	2.505			
4	2	20	81	0.00037	3.459	2.841			
4	4	50	625	0.0024	5.678	5.145			
6	2	20	729	0.0011	4.043	3.333			
8	2	50	6561	0.0093	6.941	6.505			
10	2	50	59049	0.091	7.143	6.583			
				2 (LP problems)			**3 (Linear eqs)**		
2	2	5	9	0.00020	0.976	0.840	0.000069	0.981	0.866
2	6	10	49	0.0025	1.695	1.536	0.00031	1.677	1.533
2	10	20	121	0.023	2.543	2.383	0.00074	2.511	2.410
4	2	20	81	0.0082	2.847	2.690	0.0012	2.797	2.659
4	4	50	625	2.82	5.056	4.963	0.0093	5.045	4.880
6	2	20	729	4.48	3.403	3.292	0.016	3.353	3.201
8	2	50	6561		greater than		0.24	6.291	6.129
10	2	50	59049		1 minute		3.43	6.503	6.371
				4 (min BCs)			**5 (min gradients)**		
2	2	5	9	0.000085	1.147	0.905	0.00011	0.961	0.885
2	6	10	49	0.00031	4.914	3.165	0.00044	1.910	1.514
2	10	20	121	0.00090	11.49	8.175	0.0012	3.014	2.514
4	2	20	81	0.0012	4.797	4.609	0.0015	3.199	2.766
4	4	50	625	0.0088	14.05	14.91	0.011	5.940	5.843
6	2	20	729	0.015	5.921	5.921	0.017	3.687	3.453
8	2	50	6561	0.21	14.33	15.41	0.24	7.360	7.313
10	2	50	59049	2.69	17.11	19.84	3.11	7.680	7.966

polynomials (3) in n variables with degree $l = (D, \ldots, D)^T$ and k non-zero terms. The non-zero coefficients were randomly generated with $a_i \in [-1, 1]$.

Table 1 lists the results for different values of n, D, and k; $(D+1)^n$ is the number of Bernstein coefficients. In each case 100 random polynomials were generated and the mean computation time and error are given. The results were produced with C++ on a 2.4 GHz PC. Method 2 utilizes the linear programming solver LP_SOLVE [1].

The time required for the computation of the Bernstein coefficients is included; this is equal to the time for Method 1 (constant bound functions). An upper bound on the discrepancy between the polynomial and its lower bound function over I is computed according to Theorem 4 as

$$\delta = \max_i \left\{ b_i - c_n \left(\frac{i}{l} \right) \right\}.$$

The error bounds for the bound functions resulting from application of the equilibriation transformation are labelled δ_E and are computed identically. Note that after application of the equilibriation transformation, Method 1 delivers an affine function instead of a constant.

The mean δ values for Methods 2 and 3 are very similar, with Method 3 exhibiting a slight improvement in all but the first case. The poor mean δ values for Methods 4 and 5 are greatly skewed by a small minority of cases where the shape information is incorrect. These methods are unreliable. However for any given individual polynomial, any one method may deliver a significantly superior bound function to the other, with the results only frequently identical in the $n = 2$ case. The equilibriation transformation is effective in reducing the mean error bound in almost all cases, i.e. typically $\delta > \delta_E$. For $n \leq 4$ the computation time for Methods 3-5 is of the same order of magnitude as for Method 1 (constant bound function), and is faster by orders of magnitude than Method 2. Under that method, one can typically compute bound functions in less than a second only for $n \leq 4$; for Methods 3-5 this can be done for $n \leq 8$.

5   Conclusions

We have presented several methods for the computation of affine lower bound functions for multivariate polynomials based on Bernstein expansion. A simple constant bound function based on the minimum Bernstein coefficient (Method 1) can be computed cheaply, but performs poorly. It is possible to improve this by exploiting the valuable shape information inherent in the Bernstein coefficients. With Methods 4 and 5, we have demonstrated that a naive attempt to derive such shape information based on simple differences and gradients is unreliable. Methods 2 and 3 do this reliably and in general deliver a better quality bound function. The principal difference between these two lies in the computational complexity; the general construction of Method 2 requires the solution of a linear programming problem, whereas affine bound functions according to Method 3 can be computed much more cheaply, and may therefore be of greater practical use. Indeed one may compute up to 2^n of these bound functions for a single given polynomial which jointly bound the convex hull of the control points much more closely than a single bound function from Method 2, in less time. Method 3 is therefore our current method of choice.

Methods 1-5 are limited by focussing on the shape information provided by a small number of designated control points, especially the minimum. Their performance can therefore be improved by incorporating the wider shape information provided by a broad spread of the control points. Our currently best overall results are thus obtained by combining Method 3 with the equilibriation transformation given in Section 3.5.

A fundamental limitation of our approach remains the exponential growth of the number of underlying Bernstein coefficients with respect to the number of variables. This means that many-variate (12 variables or more) polynomials cannot currently be handled in reasonable time. Future work will seek to address this limitation.

We have implemented the use of affine lower bound functions in a branch and bound framework for solving constrained global optimization problems involving a polynomial objective function and polynomial constraint functions. Relaxations based on these bound functions lead to linear programs. In practical problems, quite often only a few variables appear in the objective function and in each constraint. In this case, Method 3 may be highly suitable. If validated results are required, the solution of the linear program must be verified. This can be accomplished by using the results of [12], [16], [17].

References

1. Berkelaar M., **LP_SOLVE**: Linear Programming Code.
   ftp://ftp.ics.ele.tue.nl/pub/lp_solve/
2. Berchtold J. (2000), "The Bernstein Form in Set-Theoretical Geometric Modelling," PhD thesis, University of Bath.
3. Borradaile G. and Van Hentenryck P. (2004), "Safe and tight linear estimators for global optimization," *Mathematical Programming*, to appear.
4. Cargo G. T. and Shisha O. (1966), "The Bernstein form of a polynomial," *J. Res. Nat. Bur. Standards Vol. 70B*, 79–81.
5. Floudas C. A. (2000), "Deterministic Global Optimization: Theory, Methods, and Applications," *Series Nonconvex Optimization and its Applications Vol. 37*, Kluwer Acad. Publ., Dordrecht, Boston, London.
6. Garloff J. (1986), "Convergent bounds for the range of multivariate polynomials," *Interval Mathematics 1985*, K. Nickel, editor, *Lecture Notes in Computer Science Vol. 212*, Springer, Berlin, 37–56.
7. Garloff J., Jansson C. and Smith A. P. (2003), "Lower bound functions for polynomials," *J. Computational and Applied Mathematics Vol. 157*, 207–225.
8. Garloff J., Jansson C. and Smith A. P. (2003), "Inclusion isotonicity of convex-concave extensions for polynomials based on Bernstein expansion," *Computing Vol. 70*, 111-119.
9. Garloff J. and Smith A. P. (2004), "An improved method for the computation of affine lower bound functions for polynomials," in "Frontiers in Global Optimization," Floudas C. A. and Pardalos P. M., Eds., *Series Nonconvex Optimization with its Applications Vol. 74*, Kluwer Acad. Publ., Dordrecht, Boston, London, 135–144.
10. Hansen E. R. (1992), "Global Optimization Using Interval Analysis," Marcel Dekker, Inc., New York.
11. Hongthong S. and Kearfott R. B. (2004), "Rigorous linear overestimators and underestimators," submitted to *Mathematical Programming B*.
12. Jansson, C. (2004), "Rigorous lower and upper bounds in linear programming," *SIAM J. Optim. Vol. 14 (3)*, 914–935.
13. Kearfott R. B. (1996), "Rigorous Global Search: Continuous Problems," *Series Nonconvex Optimization and its Applications Vol. 13*, Kluwer Acad. Publ., Dordrecht, Boston, London.
14. Kearfott R. B. (2004), "Empirical comparisons of linear relaxations and alternate techniques in validated deterministic global optimization," submitted to *Optimization Methods and Software*.

15. Meyer C. A. and Floudas C. A. (2004), "Trilinear monomials with mixed sign domains: facets of the convex and concave envelopes," *Journal of Global Optimization Vol. 29 (2)*, 125–155.
16. Michel C., Lebbah, Y. and Rueher M. (2003), "Safe embeddings of the simplex algorithm in a CSP framework," in Proc. 5th Int. Workshop on Integration of AI and OR Techniques in Constraint Programming for Combinatorial Optimization Problems (CPAIOR 2003), Université de Montréal, 210–210.
17. Neumaier A. and Shcherbina O. (2004), "Safe bounds in linear and mixed-integer programming," *Math. Programming A Vol. 99*, 283–296.
18. Prautzsch H., Boehm W. and Paluszny M. (2002), "Bézier and B-Spline Techniques," Springer, Berlin, Heidelberg.
19. Rokne J. (1977), "Bounds for an interval polynomial," *Computing Vol. 18*, 225–240.
20. Ratschek H. and Rokne J. (1988), "New Computer Methods for Global Optimization," Ellis Horwood Ltd., Chichester.
21. Tawarmalani M. and Sahinidis N. V. (2002), "Convexification and Global Optimization in Continuous and Mixed-Integer Nonlinear Programming: Theory, Algorithms, Software, and Applications," *Series Nonconvex Optimization and its Applications Vol. 65*, Kluwer Acad. Publ., Dordrecht, Boston, London.
22. Zettler M. and Garloff J. (1998), "Robustness analysis of polynomials with polynomial parameter dependency using Bernstein expansion," *IEEE Trans. Automat. Contr. Vol. 43*, 425–431.

Using a Cooperative Solving Approach to Global Optimization Problems

Alexander Kleymenov and Alexander Semenov

Institute of Informatics Systems of the Russian Academy of Sciences,
pr.ac. Lavrentieva, 6 Novosibirsk, Russia, 630090
`kleymenov@ngs.ru, semenov@iis.nsk.su`

Abstract. This paper considers the use of cooperative solvers for solving global optimization problems. We present the cooperative solver Sibcalc and show how it can be used in solution of different optimization problems. Several examples of applying the Sibcalc solver for solving optimization problems are given.

Keywords: cooperative solvers, interval mathematics, interval constraint programming, global optimizations, distributive computations.

1   Introduction

Global optimization problems arise in a huge number of different applications, so the methods of effective solution of these problems are out of question. At present, there are effective algorithms for solving linear optimization problems of large dimensions, integer and quadratic optimizations, and some classes of nonlinear problems. However, solution of general nonlinear problems of large dimensions including constrained problems and mixed problems still remains a very difficult task. Therefore, the development of new approaches to solving such problems is very important and a lot of efforts are applied in this direction. Recently, new approaches based on combination of classical optimization algorithms and methods of interval mathematics and constraint programming were successfully applied to real nonlinear optimization problems. Two last classes of these methods use splitting of the search space into subspaces with further processing the resulting subspaces and pruning the subspaces without solutions. The strategies of splitting and processing determine the overall efficiency of the algorithms. It is clear that one can use parallel or distributive computations to speed up these processes. In this paper we suggest a cooperative solving approach that allows us to organize different computational schemes flexibly (in particular, different parallel schemes) and consider possible ways of using the approach.

One of the approaches to organize such computations is cooperative solving of problems by different methods and solvers. Cooperative interaction is a process of mutual solution of different parts (intersecting in general case) of the initial problem by different methods where each method provides the results

C. Jermann et al. (Eds.): COCOS 2003, LNCS 3478, pp. 86–100, 2005.

of its calculation to other methods. There are several approaches to solver cooperation and some of them are in the field of constraint programming. The following works that offer languages and environments for cooperative problem solving can be considered as the most relevant for our purposes: [5], [8], [13], [14] and [21].

The paper is organized as follows. Section 2 gives a brief comparison of our approach with other works, considers our technology for constructing cooperative solvers and the cooperative solver Sibcalc [10], the kernel of our approach. In section 3 we describe how the solver can be applied to solve optimization problems. In section 4 some numerical experiments are presented and future works are discussed in Conclusion.

2   SibCalc – An Environment for Building Cooperative Solvers

At early 90th in Novosibirsk on the basis of the interval constraint propagation and interval mathematics methods the UniCalc solver [2] has been developed. It was an integrated environment with an embedded set of methods that allowed us to solve the problems of a wide spectrum, but the solver itself was not a system opened for new methods and for organization of their cooperative interaction. But practical usage of UniCalc gave us an experience in application of such solvers. As a result a general idea of how a new solver, of a similar purpose but with an extended set of functionalities, should be developed. Based on this idea the SibCalc solver has been developed. The solver has a wide set of embedded methods and an architecture opened for new methods that allows us to use different sets of methods in problem solving. On the base of SibCalc we started the works on investigation of how to organize the cooperative problem solution and how to build an environment for specification of the cooperation means and organization of the cooperative computations. As a result, we have proposed the GMACS architecture [17] for cooperative solvers and further, on its basis, formulated an approach presented in this paper.

From all above mentioned, our main goal in the development of the ways to cooperative problem solving was organization of joint work of the existing methods, as well as easy connection of new ones to our scheme. In our approach, a cooperative solving does not necessarily mean a simultaneous running of methods, it is also allowed to run them sequentially. For example, from the source model we may first choose a linear subsystem to be solved by the interval mathematics methods, and the intervals obtained for variable values are used in solving the remaining nonlinear subsystem. Note that in our approach each of methods can use its own internal parallelization.

When comparing our approach with those mentioned above, we can say that the closest works are the environments Mosel [5] and DICE [21]. Similarities to Mosel are in the possibility to use the available methods and solvers, as well as to create new algorithms of computations in the interaction language and to place them in the library. But Mosel is inferior to our approach, since it is oriented only to optimization problems and has no means for specification of parallel and distributed computations, which is a highly essential feature of our approach.

A model of cooperative computations IWIM [1], on which DICE is based, is very close to our model in ideology. Both define the notions of channels, ports, processes (in our model, a process is a method or a solver) and anonymous messages and provide the possibility to define various kinds of cooperative interaction (asynchronous, with synchronization, etc.). At the same time, DICE, like above mentioned projects, is oriented to organization of cooperative interaction between methods on the basis of constraint programming, whereas in our approach we consider interaction of different methods and solvers, classical or based on the interval mathematics, constraint propagation, symbolic transformations, etc. It is clear that some approaches from [8], [13], [14], [21] can also be implemented in our model, which can result in construction of a hierarchical cooperative model. At the same time, the approach of [8] in our environment is completely identical to the previous one, since all solvers considered in the paper are represented by one solver in our implementation.

We have developed an environment for creation of specialized cooperative solvers for different classes of problems. The environment contains a modelling language to formulate problems, means of description of architectures of cooperative solvers (which include an interaction language to describe scenarios of computations), means of method communication, a calculation kernel, components of graphical user interfaces, etc.

Here we describe only the parts of the environment related to the topic of our paper.

2.1   SibCalc Model of Cooperative Solving

The basic concepts of our model are methods, ports, channels and calculation schemes.

A *method* in our model can be treated like a process in the IWIM model. There are one selected method that is the manager of computations and several worker methods. Worker methods can communicate and it is the responsibility of the manager method to coordinate these communications. Communications between worker methods are anonymous, that is a method does not know who it communicates with.

Methods receive and send information from or to other methods through their input and output *ports*. Each method can have several input and output ports that are used to exchange information in one direction. Ports can admit any types of information. To interconnect the ports of methods, *channels* are used. A channel connects an input port of one method with an output port of another method. A channel may implement different types of data storage and data transfer - a queue, a stack, a buffer, etc. Channels can also encapsulate some network transmission protocol that automatically allows transfer of a cooperative solver to a distributed architecture. The presence of channels allows methods to receive and send both problem data and control commands. It is known that using any of the existing software for methods communication over channels is very costly and can diminish all advantages of such computations. To avoid this problem, we have implemented our own communication protocols that significantly speeded up the process of data transferring over channels.

A cooperative interaction between methods is described by a *calculation scheme* - a digraph whose nodes are methods and arcs are the data flows. A calculation scheme completely defines the process of problem solving. It contains the information about the set of applied methods, the order of their launching, ways and directions of the data flows. In essence, a calculation scheme describes where the input information is got, the way of its transformation and what is to be received as the output. The description of a calculation scheme defines the type of cooperative interactions between the methods.

2.2 Components of the Environment for Creating Cooperative Solvers

The architecture of our environment for creating cooperative solvers is presented in Fig. 1. The main components of the environment are:

- a calculation kernel;
- a module for description of mathematical models;
- a module for construction and execution of cooperative solvers.

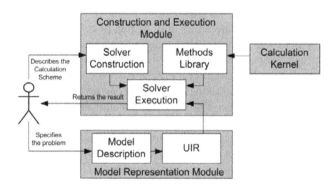

Fig. 1. The architecture of the environment

The calculation kernel contains a library of methods and applications used for problem solving. At present, the calculation kernel of the environment is represented by the SibCalc library. The SibCalc library of methods includes:

- constraint programming algorithms over finite domains (AC-4, AC-5) [18];
- constraint programming algorithms over continuous domains;
- the Newton interval method;
- methods for solving systems of interval linear equations;
- an interval linear programming method (based on of the interior point method);

- a large set of methods for search over continuous and finite domains;
- automatic and symbolic differentiation;
- a number of specialized methods for different areas of application.

The module for description of mathematical models gives the possibility to describe a mathematical model in the form very close to the conventional mathematical notation. For this purpose a special modeling language is provided that has a wide set of types, control structures and other facilities that allow us to describe any complex model in the declarative style. The language describes a model but does not require defining the method(s) of solving as other modelling languages do (AMPL [6], OPL [20]). With the help of a compiler the declarative description of the model is transformed to the universal internal representation (UIR) that is a kind of an attributed tree. This representation is used by all SibCalc methods as the input and output data. It is possible to store a model in this representation and to work later with it without repeated compilation of the source model.

The module for construction and execution of cooperative solvers provides the following facilities:

- A uniform interface which should be implemented for methods taking part in cooperative interaction. It provides a common mechanism to organize cooperative interaction between the methods and to execute them. The methods implementing this interface constitute the methods library of the environment.
- A language support of implementation of new methods - both from scratch and on the basis of the library methods.
- Channels which connect the input and output ports of methods and provide interactions between then and their control.
- The mechanism that allows us to describe the scheme of methods interaction. Such a scheme is a description of a cooperative solver and can be considered as a new method and used alongside with other methods as a component of another cooperative solver. This mechanism allows creating of hierarchical cooperative solvers.
- The mechanism that allows us to launch a cooperative solver, to arrange methods on the nodes of a heterogeneous cluster, and to control the calculation interactively. This mechanism represents a description of the cooperative solver behavior.

It should be noted here that we use Python [11] as an interaction language to describe calculation schemes and new methods as well as a language for implementation of interaction means and interfaces. The choice of Python is motivated by its power, multifunctionality and availability on different platforms. Almost all methods of the kernel library are written in C++ and also can be used on different platforms.

2.3   Technology of Cooperative Solver Construction

Using the above description of the main components of our environment, we consider the general scheme of the cooperative solver construction.

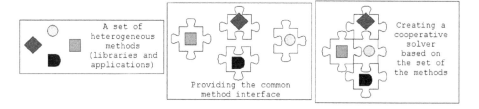

Fig. 2. The process of cooperative solver creating

When a user needs to calculate a mathematical model, he first formulates it in a high-level specialized modeling language. Then he works out a strategy of computing this model that comprises the following stages:

- decomposition of the mathematical model into components that can be solved by separate methods of the library and/or by other applications;
- search for the most efficient computational means that can solve each part of the problem;
- development of a scheme of cooperative interaction between computational means (creation of a computational scheme).

After the strategy of solving the problem is chosen, we select the methods intended to be used in cooperative solving of the problem. For each computational tool which is necessary to solve the problem and not belonging to the current library of methods, we implement a uniform method interface. A solver is being constructed on the basis of the chosen methods and the elaborated scheme of their cooperative interaction via mechanisms provided by the module for construction and execution. The Python language is used to implement this stage. Fig. 2 shows schematically the process of creating of a cooperative solver from a set of methods.

The next stage is debugging of the cooperative solver. Here, the strategy of computation of the mathematical model is improved in order to attain the maximum efficiency when solving it. After that the solver is alienated from the environment and is ready for practical use as a standalone tool.

3   Our Approach to Solution of Optimization Problems

There exist different classes of algorithms for solution of global optimization problems. We divide them into three groups:

- Classical approaches that use common methods [9].
- Methods based on the algorithms of interval mathematics [3],[7].
- Methods that use a combination of interval and constraint programming methods [4], [19].

The ways to increase efficiency in each of these groups are different. In particular, for the first group, efficiency of computation can be increased if parallel computation is organized on the basis of the algorithm properties (e.g., iterative projection algorithms) or the problem structure (model decomposition, barrier methods, etc.).

When optimization problems are solved by the methods of interval mathematics, the main efforts are applied to processing of subdomains of the initial domain so that either to find the global optimum in a subdomain, or to prune it as not containing any solution. In fact, this is attained by using various modifications of the branch-and-bound method. The approaches to efficiency increase of this method, as well as the interval algorithms used in its modifications, are studied in many papers which show us that it is possible to achieve a rather high efficiency of the algorithm as a whole and to obtain a guaranteed estimate for the global optimum at the same time.

Recently, the algorithms based on constraint programming are more actively used for solution of global optimization problems. These methods can be applied to problems over finite domains (integer and combinatorial programming) and continuous domains, as well. In the latter case, they are combined with the interval mathematics methods. To increase efficiency of the combined methods, we can apply their parallelized versions.

The computational kernel of our environment contains a number of efficient methods for solving nonlinear systems and optimization problems with variables of different types. In particular, we have implemented classical and interval-based methods of linear and nonlinear optimization, several modifications of the branch-and-bound method, and algorithms for solving the integer and mixed optimization problems. All these methods can be independently applied to solving problems of certain classes. But we think that more advantage can be gained from their joint work with the use of our cooperative model, for example, using distributed computation when different methods are running on separate computers. Information exchange during the computation essentially accelerates the process of problem solving. Each method in turn can use its own internal parallelism. It is also possible to arrange a "sequential" cooperativity, when specialized methods work one by one and pass their results to each other. We believe that joint usage of different methods allows us to solve mixed problems efficiently.

4   Numerical Experiments

Below we consider three examples of optimization problem solving by different cooperative solvers constructed by the technology described in Section 2. The first example shows the possibility of joint application of interval constraint propagation and bisection for solving optimization problems. The cooperative solver in the second example shows sequential cooperative interaction between solvers from the first example. The third example shows the distributed solution of an integer optimization problem running on several computers in parallel. It should be noted that we have just started to study how our approach can be applied to optimization problem solving, so the examples can be considered too simple. However, they show that our approach is rather efficient.

4.1   Example 1. Optimization Problem Solver Constructed on the Basis of the Interval Method of Constraint Propagation

In this example we apply our original strategy of solving optimization problems by splitting the function range [16]. The main idea of the algorithm is to find bounds of the objective function with the help of interval mathematics, split this interval into two subintervals and assign the left subinterval to the objective. Then we solve this new problem by an interval constraint propagation algorithm to check consistency of all constraints for the current domains of all variables of the problem. If the algorithm reveal inconsistency of the problem, we assign the right subinterval to the objective and repeat solving. If the problem is consistent, we usually find narrower domains of the independent variables and continue the process with these new domains until the width of the function range is less than the required accuracy. Then we apply a branch-and-bound algorithm (or bisection of the variable domains) to find the precise values of the independent variables. If we have got inconsistency for both subintervals, we backtrack to the subintervals of higher levels and repeat the process.

This algorithm can be implemented by the solver that is shown in Fig. 3. **CP** and **BISECT** are methods from the SibCalc library. **CP** implements the method of interval constraint propagation and **BISECT** organizes a backtracking search with repeated splitting over the problem variables.

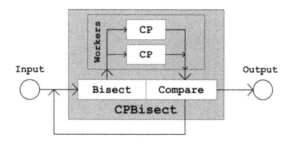

Fig. 3. An example of Bisect and CP cooperation

In this solver the source model is at the input of the **Bisect** method that divides the range of the objective function in two parts and passes them to two **CP** methods working in parallel. The results of the **CP** methods are compared and the model containing an interval with the minimal value of the objective is chosen. If the width of the interval does not satisfy the specified accuracy, the domain is again split into two parts until the accuracy is reached.

The solver constructed by the above computational scheme is saved in the library of methods as a new **CPBisect** method which can be further used in constructing other solvers. The Python script that describes the CPBisect computational scheme is as follows:

```
from scf.method import method
from scf.methods.cp import CP
from scf.methods.bisect import Bisect

class CPBisect(method):
    def __init__(self, target, acc=1e-6, *args, **kwargs):
        method.__init__(self, *args, **kwargs)

        self.target = target
        self.acc = acc

        self.bisect = bisect = Bisect(target, acc=acc, \
                                 name="%s->Bisect" % self.name)
        self.cp_1 = cp_1 = CP(name="%s->CP-1" % self.name)
        self.cp_2 = cp_2 = CP(name="%s->CP-2" % self.name)

        bisect.to_workers.connect(cp_1.input)
        bisect.to_workers.connect(cp_2.input)
        bisect.from_workers.connect(cp_1.output)
        bisect.from_workers.connect(cp_2.output)

        bisect.input.connect(self.input)
        bisect.output.connect(self.output)

    def run(self):
        self.bisect.start()
        self.cp_1.start()
        self.cp_2.start()

        while self.works:
            time.sleep(0.01)

        self.bisect.stop()
        self.cp_1.stop()
        self.cp_2.stop()
```

In the first example we need to find the global optimum of a real function whose parameters are bound by a set of additional constraints given by equalities and inequalities. The problem was suggested by Janos Pinter [15] as a test for SibCalc. The problem description below is in the SibCalc modelling language.

```
real x1, x2, x3, x4, goal;
/* minimize goal */
goal = x1^2 + 2.*x2^2 + 3.*x3^2 + 4.*x4^2;

/* Constrained to: */
x1^2 + 3.*x2^2 - 4.*x3*x4 - 1. = 0.;
x4 = max(x4, 1.e-4);
sqrt(5.*abs(x1/x4)) + sin(x2) - 1. = 0.;
```

```
(x1 - x2 - x3 + x4)^2 - 1.=0;
x2*x3 + (x1 - sin(x4))^2 - 3.=0;

-5 <= x1; x1 <= 3;
-3 <= x2; x2 <= 3;
-5 <= x3; x3 <= 4;
-6 <= x4; x4 <= 2;
```

To run the CPBisect method for this model that is in file example1.slv, the following Python script can be used:

```
import time
from pysolver.model import model
from scf.methods import CPBisect

model = model("example1.slv")
cp_bisect = CPBisect(target="goal", acc=1e-5)
beg = time.time()

cp_bisect.input(model)
cp_bisect.run()

print "Result:\n", cp_bisect.output()
print "Time:", time.time() - beg
```

Applying the **CPBisect** method with accuracy 10^{-4}, it takes 224.102 seconds on the Athlon-1700Mhz processor to find the following solution:

```
x1 = [-0.8526256398303158, -0.8525175683959241];
x2 = [-0.8883181498321493, -0.8880895602332961];
x3 = [ 0.3871006162671068,  0.387350031063668];
x4 = [ 1.351541041416323,   1.351661572804863];
goal=[10.06104506101511,   10.0626879996752];
```

However, solving this problem with accuracy 10^{-6} requires 28062.6 seconds that is unacceptable for real applications. The next example shows how it is possible to reduce the calculation time and get results with a higher accuracy.

4.2   Example 2. A Solver That Implements the Sequential Cooperativity with Varying Accuracy of Computations

The goal of this example is to show how the sequential cooperativity can help to speed up the calculations. Here, we search for the minimum of the function `goal` from Example 1 with the accuracy 10^{-10}. To reduce the number of splittings, we will increase the accuracy step-by-step. Each next step will take the reduced domain from the previous step and bisect it with a higher accuracy. As a result, such a solver can be described by the computational scheme from Fig. 4.

Note that in this example we build a cooperative solver on the basis of the **CPBisect** solver from Example 1. The newly built solver also allows its further reusage as a new method. The computational scheme from Fig. 4 corresponds to the following method described in the internal language:

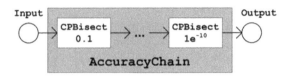

Fig. 4. An example of a sequential cooperation

```
from scf.methods import method, CPBisect

class AccuracyChain(method):
    def __init__(self, target, acc=1e-5, *args, **kwargs):
        method.__init__(self, *args, **kwargs)
        self.target = target
        self.accuracy = acc
        self.direction = dir
        self.solvers = []

        accuracy = 0.999999999999
        channel = self.input
        while accuracy > self.accuracy * 0.1:
            bisect = CPBisect(name="CPBisect<%f>" % accuracy, \
                              target=self.target, acc=accuracy)
            channel.connect(bisect.input)
            channel = bisect.output
            self.solvers.append(bisect)
            accuracy *= 0.1
        channel.connect(self.output)

    def run(self):
        for s in self.solvers:
            s.start()

        while not self.output.has_data():
            time.sleep(0.1)

        for s in self.solvers:
            s.stop()

        self.works = 0
```

This description shows how to build a solver which allows its further reusage. Creation of a solver copy, its initialization, launch and output of a result should be made as follows:

```
import time
from pysolver.model import model
from scf.methods import AccuracyChain
```

```
model = model("example1.slv")
acc_chain = AccuracyChain(target="goal", acc=1e-10)
beg = time.time()

acc_chain.input(model)
acc_chain.run()

print "Time:", time.time() - beg
print "Result:\n", acc_chain.output()
```

The efficiency of this approach is proved by comparison of the calculation times with Example 1. The running time of computations with the gradually increasing accuracy using the solver above is only 1.5 seconds. The results are as follows:

```
x1 = [-0.8525645916123212,  -0.852564591516835];
x2 = [-0.8881772432353092,  -0.8881772430316305];
x3 = [ 0.3872150020694321,   0.3872150022375183];
x4 = [ 1.351602350526265,    1.351602350686664];
goal=[10.06170604312326,    10.06170604446985];
```

4.3 Example 3. A Solver That Implements the Distributed Asynchronous Interaction Between Several Methods (Distributed Branch-and-Bound)

Let us consider an example of a cooperative solver which implements the distributive branch-and-bound method **DBB**. Taking into account the fact that this is a search method, we can substantially accelerate the process of solving large optimization problems by making computations distributed. The computational scheme of the cooperative solver that implements the distributed branch-and-bound method is represented in fig. 5.

The computations involve many copies of the branch-and-bound method controlled by one distinguished method. In the search within its domain, this method periodically asks the system if there is a free computational station. If so, it creates a new copy of the DBB method and passes its current search domain as the input data of this copy. One of the possibilities provided by our environment is the possibility to dynamically scale the computational cluster. Thus, if we are in lack of computational capacity of a cluster, we can supply it with new computational stations without a pause in the process of solution.

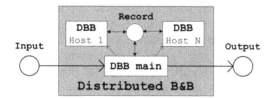

Fig. 5. The distributed branch-and-bound method

It should be stressed that the main feature of the cooperative solver is the possibility to dynamically change its configuration in the process of solving a problem. In addition, this example shows the possibility of the distributed shared data processing. This mechanism is implemented with a channel **record** writing to which is synchronized (the *data output consistency* model).

The solver here presented has been applied to the traveling salesman problem for various dimensions on the clusters of different computational capacity. Below we give the results of several experiments on Ultra-Sparc1 (450 Mhz).

Dimension	Number of computers		
	1	2	3
12	17.93	10.62	6.46
15	77.94	42.04	28.12
20	1403.79	684.73	497.23

5   Conclusion and Future Work

This paper describes the use of the cooperative approach to solution of optimization problems. The architecture for cooperative solver construction here proposed differs from the common ones. In particular, we do not restrict the spectrum of methods that can be used for solver constructing and provide a user with a possibility to implement any computational scheme. There are synchronization channels and channels for transmission of the control signals, which allows us to implement mixed asynchronous-synchronous schemes. The potentialities for constructing the distributed solvers based on this architecture allow one to use powerful computational capacity of the cluster systems, as well as local networks. Summing up, we can state that the key points of our approach are the ability to organize different computational schemes and flexible means for organizing these schemes.

As it was mentioned above, we have employed Python [11] as an interactive language. This language makes it easy to embed new computational libraries into the system when reusing existing components. Python have also been used in implementation of our own data exchange protocols and this made it possible to significantly increase efficiency of distributed computations. When we embedded SibCalc as the computational kernel of the environment, we have implemented the Python interface for the whole library of the solver. The environment can be extended with other applications in a similar way.

To increase efficiency and to build practical applications, it is necessary to conduct further investigations, experiments and projects. This relates to supplement of the library with new optimization methods and parallelization of the existing methods of the library, as well as to creation of computational schemes that efficiently use these methods. At present, the library of methods comprises only the methods that were developed for the computational kernel of the Sib-Calc solver but there are no methods implemented as computational schemes for cooperative problem solving.

On this basis, our plans for future consist of the following tasks:

- Extension of the set of methods used in solution of optimization problems.
- Creation of specialized methods for solution of global optimization problems and computational schemes based on these methods.
- Development of cooperative parallel solvers for optimization problems.
- Development of specialized solvers for mixed optimization problems.

References

1. Arbab, F.: The IWIM Model for Coordination of Concurrent Activities. Proc. of First International Conference COORDINATION'96. Cesena, Italy, LNCS 1061, (1996) 34-56
2. Babichev A., et al.: UniCalc, A Novel Approach to Solving Systems of Algebraic Equations. Interval Computations, **2** (1993) 29-47
3. Alefeld, G., Herzberger, Ju.: Introduction in Interval Computations. Academic Press (1983)
4. Notes of the 1st International Workshop on Global Constrained Optimizations and Constraint satisfaction. Valbonne - Sophia Antipolis, France, October (2002)
5. Colombani, Y., Heipcke, S.: Combining Solvers and Solutions Algorithms with Mosel. Proc. of the 4th International Workshop on Integration of AI and OR Techniques in Constraint Programming for Combinatorial Optimisation Problems CP-AI-OR'02. (2002) 277-290
6. Fourer, R., Gay, D., Kernighan, B.: AMPL: A Modeling Language for Mathematical Programming. The Scientic Press (1993)
7. Hansen, E.: Global Optimization Using Interval Analysis. Marcel Dekker (1992)
8. Hofstedt, P., Seifert D., Godehardt E.: A Framework for Cooperating Solvers - A Prototypic Implementation. Proc. of CoSolv'01 workshop. Paphos, Cyprus, December (2001) 11-25
9. Horst, R., Hoang, T.: Global Optimization: Deterministic Approaches. Springer Verlag (1996)
10. Kleymenov, A., Petunin, D., Semenov, A., Vazhev, I.: A Model of Cooperative Solvers for Computational Problems. Proc. 4th Int. Conference PPAM'01, Naleczow, Poland, LNCS 2328, (2002) 797-802
11. Lutz, M.: Programming Python. 2nd Edition. O'REILLY & Associates, Inc. (2001)
12. Marti, P., Rueher, M.: A Distributed Cooperating Constraint Solving System. International Journal on Artificial Intelligence Tools, **4** (1995) 93-113.
13. Monfroy, E.: The Constraint Solving Collaboration in BALI. Proc. of the International Workshop Frontiers of combining systems FroCoS'98. (1998)
14. Pajot, B., Monfroy, E.: Separating Search and Strategy in Solver Cooperations. Proc. of the 5th International Conference "Perspectives of System Informatics" (PSI'03). Novosibirsk, Russia, July (2003) 275-281
15. http://is.dal.ca/ jdpinter/index.html
16. Semenov, A.: Solving Integer/Real Nonlinear Equations by Constraint Propagation. Technical Report N12, Institute of Mathematical Modelling, The Technical University of Denmark, Lyngby, Denmark, (1994)
17. Semenov, A., Petunin, D., Kleymenov, A.: GMACS - the General-Purpose Module Architecture for Building Cooperative Solvers. Proc. ERCIM/Compulog Net Workshop on Constraints, Padova, Italy (2000)

18. Tsang, E.: Foundations of Constraint Satisfaction. Academic Press, Essex (1993)
19. Van Hentenryck, P., Michel, L., Deville, Y.: Numerica: A Modelling Language for Global Optimization. MIT Press (1997)
20. Van Hentenryck, P.: The OPL Optimization Programming Language. MIT Press (1999)
21. Zoeteweij, P.: A Coordination-Based Framework for Distributed Constraint Solving. Proc. ERCIM/CologNet Workshop on Constraint Solving and Constraint Logic Programming, Cork, Ireland, LNAI 2627 (2003) 171-184

Global Optimization of Convex Multiplicative Programs by Duality Theory

Rúbia M. Oliveira and Paulo A.V. Ferreira*

University of Campinas,
Faculty of Electrical & Computer Engineering,
13084-970 Campinas/SP Brazil
{rubia, valente}@dt.fee.unicamp.br

Abstract. A global optimization approach for convex multiplicative problems based on the generalized Benders decomposition is proposed. A suitable representation of the multiplicative problem in the outcome space reduces its global solution to the solution of a sequence of quasi-concave minimizations over polytopes. Some similarities between convex multiplicative and convex multiobjective programming become evident through the methodology proposed. The algorithm is easily implemented; its performance is illustrated by a test problem.

Keywords: Global optimization, convex multiplicative programming, multiobjective programming, duality theory, numerical methods.

1   Introduction

This paper is concerned with convex multiplicative problems, a class of minimization problems involving a product of convex functions in its objective or in its constraints. Applications of multiplicative programming are found in areas such as microeconomics and geometric design [1]. An important source of multiplicative problems are certain convex multiobjective problems in which the product of the individual objectives plays the role of a surrogate objective function. A usual strategy adopted by algorithms for convex multiplicative problems is to project this (generally nonconvex) problem onto the m-dimensional real space, where m is the number of convex functions, so as to coordinate its global solution from the *outcome space* [1], [2] [3].

Projection and decomposition are well-established strategies in the mathematical programming literature [4] and their principles have been progressively extended to global nonconvex optimization problems [5], [6]. The algorithm we propose for the special class of convex multiplicative problems is inspired in a traditional projection-decomposition technique based on convex duality theory, known as generalized Benders decomposition [7]. The distinguishing feature of our algorithm is to handle the individual convex function values as *complicating*

---

* This work was partially sponsored by grants from CNPq and FAPESP, Brazil.

variables (in the terminology of [7]), in order to obtain an outer approximation of the problem in the outcome space. The solution of this relaxed problem, based on an adequate vertex enumeration procedure, is then sent to a min-max subproblem, which tests it with respect to its ϵ-feasibility. If not ϵ-feasible, the solution leads to an improved outer approximation of the original problem, whose ϵ-optimum is eventually obtained by the algorithm after finitely many iterations.

The approach adopted in this paper naturally exposes similarities between convex multiplicative and convex multiobjective programming. The implementation of the resulting algorithm is simple and preliminary experience with test problems has shown that its convergence to the ϵ-optimum is actually attained in a relatively small number of iterations.

The paper is organized as follows. In Section 2 we formulate the convex multiplicative problem and analyse its connections with convex multiobjective programming. In Section 3 a decomposition approach for convex multiplicative programming based on duality theory is proposed. Implementation and convergence issues are also discussed. A numerical example is discussed in Section 4. Conclusions are presented in Section 5.

2   A Multiobjective View of Multiplicative Problems

Multiobjective programming concepts and results [8] have implicitly provided a basis for the development of some algorithms for multiplicative programming problems [1], [2], [3]. An explicit relationship between these two fields of the mathematical programming based on the concept of *efficient solution* is presented in this section.

Consider the convex multiplicative problem

$$(P_M) \quad \left| \begin{array}{l} \text{minimize } F(f(x)) = \displaystyle\prod_{i=1}^{m} f_i(x) \\[2ex] \text{subject to} \quad g_i(x) \leq 0, \ 1, 2, \ldots, p, \end{array} \right.$$

where $f_i : \Re^n \to \Re$, $i = 1, 2, \ldots, m$ $(m \geq 2)$ and $g_j : \Re^n \to \Re$, $j = 1, 2, \ldots, p$, are continuous convex functions. As usual, we assume that

$$\Omega := \{x \in \Re^n \ : \ g_j(x) \leq 0, \ j = 1, 2, \ldots, p\}$$

is a nonempty, compact (convex) set, and that each f_i is positive over Ω. We associate to (P_M) the problem of minimizing the vector-valued objective $f := (f_1, f_2, \ldots, f_m)$ over Ω, with $F : \Re^m \to \Re$ playing the role of a special *disutility function* [8] that aggregates the individual objectives $f_1, f_2, \ldots, f_m$. Under these assumptions, $F(f(x))$ is generally nonconvex over Ω but quasiconcave over $\{f(x) \ : \ x \in \Omega\}$ [9].

Multiobjective minimization problems are comprehensively treated in [8], for example. A solution $x^* \in \Omega$ is said to be an *efficient* solution of the multiobjective (multiplicative) problem (P_M) if there exists no other $x \in \Omega$ such that (in the componentwise sense) $f(x) \le f(x^*)$ and $f(x) \ne f(x^*)$. We denote the set of all efficient solutions as effi(Ω). Given that

$$\frac{\partial F(f(x))}{\partial f_i(x)} = \prod_{j \ne i}^{m} f_j(x) > 0$$

for all $x \in \Omega$, it follows that $F(f(x))$ is increasing with respect to each $f_i(x)$, thus assuring the validity of a fundamental property derived in [10].

Proposition 1. Let $x^* \in \Omega$ be an optimal solution of the convex multiobjective (multiplicative) problem (P_M). Then $x^* \in$ effi(Ω).

It is well-known is the multiobjective programming literature [8] that $x \in \Omega$ is an efficient solution of (P_M) if and only if there exists a nonnegative vector $w \in \Re^m$ such that x is also a solution of the convex *weighting problem*

$$(P_W) \quad \left|\begin{array}{l} \text{minimize } \langle w, f(x) \rangle := \sum_{i=1}^{m} w_i f_i(x) \\[2mm] \text{subject to} \quad g_i(x) \le 0, \ 1, 2, \ldots, p. \end{array}\right.$$

Without loss of generality, it can be assumed that

$$w \in \mathcal{W} := \{ w \in \Re^m \ : \ w \ge 0, \ \sum_{i=1}^{m} w_i = 1 \}.$$

There is an obvious relationship between the weighting problem (P_W) and the following characterization (in terms of (P_W)) of the optimal solution of the multiplicative problem (P_M) [11].

Theorem 1. Let x^* be an optimal solution of (P_M). Then any optimal solution of (P_W) is optimal to (P_M) if $w = w^*$ where

$$w_i^* = \prod_{j \ne i} f_j(x^*) > 0, \ i = 1, 2, \ldots, m.$$

Incidentally $w_i^* > 0$, $i = 1, 2, \ldots, m$ is a sufficient condition for efficiency [8], that is, the optimal solution of (P_M) is surely an efficient solution of the associated multiobjective problem. However, the optimal weighting vector w^* depends on the (unknown) optimal solution of (P_M), which prevents Theorem 1 from being directly applied. What we propose in this paper can be viewed as an iterative method for obtaining w^* and thus the optimal solution of (P_M).

3   A Decomposition Approach

The use of outcome space formulations in multiobjective programming is very common [8]. Thus it is not surprising that some algorithms [1], [3] adopt the following *outcome space formulation* for convex multiplicative programs:

$$(P_y) \left| \begin{array}{l} \text{minimize } F(y) = \prod_{i=1}^{m} y_i \\ \\ \text{subject to } \quad y \in \mathcal{Y}, \end{array} \right.$$

where

$$\mathcal{Y} := \{ y \in \Re^m \; : \; y = f(x), \quad x \in \Omega \}$$

is the *outcome space*. The continuity of f and the compactness of Ω imply the compactness of $\mathcal{Y}$. The set of all efficient solutions in the outcome space is given by effi$(\mathcal{Y}) = f(\text{effi}(\Omega))$. It is readily seen that if $y \in$ effi$(\mathcal{Y})$ then $y \in \partial\mathcal{Y}$, where $\partial\mathcal{Y}$ denotes the boundary of $\mathcal{Y}$. Furthermore, $\mathcal{Y}$ admits a supporting hyperplane at each $y \in$ effi$(\mathcal{Y})$ [8], which has motivated the development of outer approximation algorithms for convex multiplicative problems.

Defining the sets $\mathcal{D} := \{ d \in \Re^m \; : \; d \geq 0 \}$, the nonnegative orthant in $\Re^m$, and $\mathcal{Y} + \mathcal{D} := \{ z \in \Re^m \; : \; z = y + d, \; y \in \mathcal{Y}, \; d \in \mathcal{D} \}$, the following statements hold [8].

Theorem 2.
a) effi$(\mathcal{Y})$ = effi$(\mathcal{Y} + \mathcal{D})$;
b) $\mathcal{Y} + \mathcal{D}$ is a convex set.

The convex set $\mathcal{Y} + \mathcal{D}$ can be explicitly represented as

$$\mathcal{F} := \{ y \in \Re^m \; : \; f(x) \leq y \quad \text{for some } x \in \Omega \},$$

given that any $y \in \mathcal{F}$ is actually a sum of elements of $\mathcal{Y}$ and $\mathcal{D}$. Theorem 2 allows us to reformulate (P_y) as a problem with a convex feasible set:

$$(P_\mathcal{F}) \left| \begin{array}{l} \text{minimize } F(y) \\ \\ \text{subject to } \quad y \in \mathcal{F}. \end{array} \right.$$

Theorem 3. *Let $y^* \in \mathcal{F}$ be an optimal solution of $(P_\mathcal{F})$. Then $y^* \in$ effi$(\mathcal{Y})$, and y^* is also an optimal solution of (P_y).*

Proof: If $y^* \in \mathcal{F}$ solves $(P_\mathcal{F})$, there exists a $x^* \in \Omega$ such that $y^* = f(x^*) \in \mathcal{Y}$. Otherwise, if $y^* \geq f(x^*)$ and $y^* \neq f(x^*)$, then $y^0 = f(x^*)$ would contradict the optimality of y^*, since $y^0 \in \mathcal{F}$ and $F(y^0) < F(y^*)$. Hence $y^0 = y^* = f(x^*)$. It is also evident that $y^* \in$ effi$(\mathcal{Y})$. Because $\mathcal{Y} \subset \mathcal{F}$, we conclude that y^* is also an optimal solution of (P_y). $\qquad\square$

A fundamental step towards the solution of $(P_{\mathcal{F}})$ is to determine whether some $y \in \mathcal{R}^m$ belongs to $\mathcal{F}$ or not. This question is answered by an important convex analysis result [7].

Theorem 4. $y \in \mathcal{F}$ *if and only if y satisfies the infinite system of linear inequalities*

$$\min_{x \in \Omega} \langle w, f(x) - y \rangle \leq 0 \quad \text{for all } w \in \mathcal{W}, \tag{1}$$

where

$$\mathcal{W} := \{w \in \mathcal{R}^m \; : \; w \geq 0, \sum_{i=1}^{m} w_i = 1\}.$$

In practice, we implement the following Corollary of Theorem 4: $y \in \mathcal{F}$ if and only if $\Theta(y) > 0$, where

$$\Theta(y) := \max_{w \in \mathcal{W}} \phi(w) \tag{2}$$

and

$$\phi(w) := \min_{x \in \Omega} \langle w, f(x) - y \rangle. \tag{3}$$

Any optimal solution of the convex minimization problem in (3) for a given $w \in \mathcal{W}$ is represented as $x(w)$. Then it is possible to show that $\xi = f(x(w)) - y$ is a *subgradient* of ϕ at $w \in \mathcal{W}$ and that an *outer approximation* procedure can be used to solve the min-max problem in (2). See [12] for details.

Algorithm A_1

Step 0: Choose $w^0 \in \mathcal{W}$ and set $l \leftarrow 0$;
Step 1: Solve the convex programming problem

$$(P_W) \; \left| \; \begin{array}{l} \text{minimize } \langle w^l, f(x) \rangle \\[2mm] \text{subject to} \quad x \in \Omega, \end{array} \right.$$

obtaining $x(w^l)$;

Step 2: Solve the linear programming problem

$$(P_L) \; \left| \; \begin{array}{l} \text{minimize } \sigma \\[2mm] \text{subject to} \quad \sigma \geq \langle w, f(x(w^i)) - y \rangle, \quad i = 0, 1, \ldots, l \\[1mm] \qquad\qquad w \in \mathcal{W}, \; \sigma \in \mathcal{R}. \end{array} \right.$$

obtaining σ^{l+1}, w^{l+1} and $\phi(w^{l+1})$. If $\sigma^{l+1} - \phi(w^{l+1}) < \epsilon_1$ where $\epsilon_1 > 0$ is a small tolerance, make $\Theta(y) = \sigma^{l+1}$ and stop. Otherwise, set $l \leftarrow l + 1$ and return to Step 1.

A second outer approximation procedure is employed to solve the multiplicative problem in the outcome space. We denote the k-th outer approximation of $\mathcal{F}$, problem $(P_\mathcal{F})$, as $\mathcal{F}^k$. An initial approximation $\mathcal{F}^0$ containing $\mathcal{F}$ can be defined as $\mathcal{F}^0 := \{y \in \Re^m : y \geq \underline{y}\}$, where $\underline{y}$ denotes the $utopian$ vector composed of the individual minima of the convex functions $f_1, f_2, \ldots, f_m$ over Ω.

Algorithm A_2

Step 0: Find $\mathcal{F}^0$ and set $k \leftarrow 0$;
Step 1: Solve the relaxed multiplicative problem

$$(P_{\mathcal{F}^k}) \left| \begin{array}{l} \text{minimize } F(y) \\[2mm] \text{subject to} \quad y \in \mathcal{F}^k, \end{array} \right.$$

obtaining y^k;

Step 2: Find $\Theta(y^k) = \langle w^k, f(x(w^k)) - y^k \rangle$ using algorithm A_1. If $\Theta(y^k) < \epsilon_2$, where $\epsilon_2 > 0$ is a small tolerance, stop: y^k solves $(P_\mathcal{F})$ and $x(w^k)$ solves (P_M). Otherwise, define

$$\mathcal{F}^{k+1} := \{y \in \mathcal{F}^k : \langle w^k, y \rangle \geq \langle w^k, f(x(w^k)) \rangle\},$$

set $k \leftarrow k+1$ and return to Step 1.

Theorem 5. *Any limit point y^* of the sequence $\{y^k\}$ generated by algorithm A_2 is an optimal solution of the convex multiplicative problem $(P_\mathcal{F})$.*

Proof: Note that problem $(P_{\mathcal{F}^k})$ always has an optimal solution; its optimal objective value is bounded below at $y = \underline{y}$. At any iteration k, the last linear inequality incorporated into $\mathcal{F}^k$ is

$$\langle w^k, y - f(x^k) \rangle \geq 0,$$

and can be rewritten as

$$\langle w^k, y - y^k \rangle \geq \langle w^k, f(x^k) - y^k \rangle,$$

$$= \Theta(y^k).$$

At any subsequent iteration $p > k$ of algorithm A_2, we must have

$$\Theta(y^k) \leq \langle w^k, y^p - y^k \rangle,$$

$$\leq \|w^k\| \, \|y^p - y^k\|,$$

$$\leq \|y^p - y^k\|,$$

because $\|w^k\| \leq 1$ for all $w^k \in \mathcal{W}$. As $k \to \infty$, we obtain $y^k \to y^*$, $y^p \to y^*$ and $\Theta(y^*) \leq 0$. Therefore, $y^* \in \mathcal{F}$, that is, y^* is a feasible solution of $(P_\mathcal{F})$. Denoting by F^* the optimal value of $(P_\mathcal{F})$, and knowing that $\mathcal{F} \subset \mathcal{F}^k$ for all $k = 0, 1, 2, \ldots$, we conclude that $F^* \geq F(y^*)$. Consequently, y^* is an optimal solution of $(P_\mathcal{F})$. $\square$

A number of practical considerations have been taken into account while implementing Algorithm A_2.

Initial Approximation of $\mathcal{F}$. We initially approximate $\mathcal{F}$ by the convex politope $\mathcal{F}^0 = \{y \in \Re^m : \underline{y} \leq y \leq \overline{y}\}$, where $\overline{y}$ is such that effi$(\mathcal{Y}) \subset \mathcal{F}^0$. The components of $\overline{y}$ can be defined as the individual maxima of the convex functions $f_1, f_2, \ldots, f_m$ over Ω, which involves the solution of another m convex maximization problems. An alternate numerical procedure for finding $\mathcal{F}^0$ has been suggested in [3].

The solution of $(P_{\mathcal{F}^k})$. Since the global minimum of any quasiconcave objective over a polytopic set is attained at a vertex of the polytope [9], the global minimum of problem $(P_{\mathcal{F}^k})$ is attained at a vertex of $\mathcal{F}^k$. We have implemented a vertex enumeration procedure based on the adjacency lists algorithm proposed in [13] (see also [9]). Given that any optimal solution of $(P_{\mathcal{F}^k})$ is necessarily an efficient vertex, only efficient vertices on the lists need to be evaluated. As an example, initially there are 2^m vertices in the list but the solution of $(P_{\mathcal{F}^0})$ is obviously $y^0 = \underline{y}$, the efficient one.

The use of Deepest Cuts. If $\Theta(y^k) < \epsilon_2$, the algorithm terminates with an ϵ_2-optimal solution of (P_M). Otherwise, the *most violated constraint* by y^k (in the sense that the left-hand side of (1) is maximized over $\mathcal{W}$) is determined and added to $\mathcal{F}^k$, significantly improving the outer approximation of $\mathcal{F}$. Thus a *deepest cut* in $\mathcal{F}^k$ is produced at each iteration. Although any violated constraint, that is, any linear inequality such that

$$\langle w^k, f(x(w^k)) - y^k \rangle > 0, \quad w^k \in \mathcal{W},$$

could be used to define $\mathcal{F}^{k+1}$, the extra effort invested in finding the most violated constraint has resulted in a faster convergence of the algorithm. By using the most violated constraints, we also limit the growth of the number of vertices in problem $(P_{\mathcal{F}^k})$;

Convergence of Algorithm A_2. The most violated constraint supports $\mathcal{F}$ at $x^k = x(w^k)$. Numerical experience has shown that $\Theta(y^k)$ is related to the infinity norm between y^k and $f(x(w^k)) \in \mathcal{F}$, which has been used to guide the selection of ϵ_2. Numerical tests with algorithm A_2 have also shown that, in general, a relatively small number of iterations (cuts in $\mathcal{F}^0$) are needed for obtaining a sufficiently good outer approximation of $\mathcal{F}$.

Computational Effort. Most of the computational effort required by algorithm A_2 is concentrated at Step 2, where $\Theta(y^k)$ is computed by algorithm A_1. While the linear programming minimizations (Step 2 of A_1) are relatively inexpensive, the nonlinear ones (Step 1 of A_1) demand some effort, although their convexity enable the use of very efficient convex programming algorithms. The codification and preparation efforts related to the approach proposed (Algorithms A_1 and A_2) seem to be small compared with other approaches available in the literature [1], [3]. The numerical results reported in the next section have been obtained with an implementation of the algorithms in MATLAB (V. 6.1)/Optimization Toolbox (V. 2.1.1) [14].

4   Numerical Examples

Consider the illustrative example discussed in [3], where an alternate algorithm for convex multiplicative problems combining branch and bound and outer approximation techniques is proposed. The data involved are: $n = m = 2$,

$$f_1(x) = (x_1 - 2)^2 + 1, \quad f_2(x) = (x_2 - 4)^2 + 1,$$
$$g_1(x) = 25x_1^2 + 4x_2^2 - 100, \quad g_2(x) = x_1 + 2x_2 - 4.$$

Letting $\underline{y} = (1, 1)$, $\bar{y} = (18, 38)$ (as in [3]), $\epsilon_1 = 0.001$ and $\epsilon_2 = 0.01$, we have obtained the results reported in Table 1.

Table 1. Convergence of Algorithm A_2

k	y^k	w^k	$x(w^k)$	$\Theta(y^k)$
0	(1.0000,1.0000)	(0.4074,0.5926)	(0.0000,2.0000)	4.0000
1	(1.0000,7.7500)	(0.6585,0.3415)	(1.3547,1.3226)	0.4170
2	(1.0000,8.9711)	(0.8129,0.1871)	(1.7014,1.1493)	0.1016
3	(1.0000,9.5139)	(0.8907,0.1093)	(1.8509,1.0745)	0.0247
4	(1.0000,9.7394)	(0.9451,0.0549)	(1.9009,1.0495)	0.0074

Algorithm A_2 has converged after only 5 iterations to the ϵ_2-global solution $x^4 = (1.9009, 1.0495)$. The optimal multiplicative function value has been $f_1(x^4)f_2(x^4) = 9.8008$. As expected, x^4 is an efficient solution for the associated convex bi-objective problem, as both components of w^4 are positive. Indeed, all the intermediate solutions generated by algorithm A_2 are efficient. With a convergence criterion equivalent to $\epsilon_2 = 0.025$, the algorithm proposed in [3] has converged after 8 iterations.

A more detailed investigation about the performance of the proposed algorithm will be carried out with basis on the following subclass of convex multiplicative problems [15]:

$$(P_q) \quad \left| \begin{array}{l} \text{minimize } (\langle c^0, x \rangle + d^0) \displaystyle\prod_{j=1}^{q} \left[\langle c^j, x \rangle + x^T \text{diag}(d_1^j, d_2^j, \ldots, d_n^j)x \right] \\[2mm] \text{subject to } \quad Ax \leq b, \quad 0 \leq x \leq \bar{x}. \end{array} \right.$$

Table 2. Results from Kuno *et al.* (1993) ($q = 2$)

n	40	60	80	100	100	120	120
m	50	50	60	80	100	100	120
AC	34.6	45.5	43.1	43.7	43.0	52.7	51.4
SD	8.62	19.41	12.51	10.63	14.72	10.74	17.60
AV	140.7	192.9	181.9	185.3	181.3	226.9	222.6
SD	40.71	99.68	63.14	50.54	70.93	51.36	87.26
AT	25.12	100.61	239.44	659.80	685.04	1268.57	1801.33
SD	25.44	71.23	88.88	532.53	303.05	680.56	1136.87

where $A \in \Re^{m \times n}$, $b \in \Re^m$, $c^j \in \Re^n$ and $d^j \in \Re^n$ ($j = 0, 1, \ldots, q$) are constant matrices with entries randomly generated in the interval $[0, 10^2]$ ($\bar{x} = 10^6$) The objective function in (P_q) is the product of one linear and q quadratic (convex) functions. Table 2 reproduces the results obtained in [15] for $q = 2$ with an outer approximation method. Ten examples have been solved for each size (n, m) of problems. The average number of cuts (AC) and vertices (AV), the average CPU time (AT, in seconds), as well as their standard deviations (SD), are indicated in Table 2.

The results obtained with Algorithm A_2 are presented in Table 3. The average number of cuts and vertices produced are significantly smaller than those generated by the method discussed in [15]. (Note that AC is the number of times that problem (P_W) is solved.) The use of most violated constraints has accelerated the convergence of the algorithm and fewer cuts has been actually needed. In addition, as AV is proportional to AC, the number of vertices produced is considerably smaller. The size (n, m) of the problem does not seem to have a significant effect on AC (and hence, on AV), whatever the method considered (Tables 2 e 3). In Table 3 we also present the average number of times that Algorithm A_1 is invoked by Algorithm A_2 (AS). The average CPU times reported have been obtained by using a personal computer (Pentium IV, 2.4GHz, 512MB RAM).

Table 3. Results with Algorithm A_2 ($q = 2$)

n	40	60	80	100	100	120	120
m	50	50	60	80	100	100	120
AC	10.75	12.55	9.60	11.40	11.30	9.75	10.05
SD	5.10	7.25	6.23	6.10	7.34	5.03	6.21
AV	48.85	58.35	41.80	49.05	50.10	39.30	42.75
SD	28.49	37.93	32.95	28.45	37.69	20.91	28.38
AT	5.67	12.93	22.17	40.89	43.87	66.62	64.71
SD	1.10	3.03	4.70	9.05	9.83	18.07	11.89
AS	4.79	4.39	5.30	4.80	4.87	5.52	5.11
SD	2.22	1.98	1.88	1.63	2.24	1.74	2.09

Table 4. Results with Algorithm A_2 $(q = 5)$

n	20	40	40	60	80
m	30	30	50	50	60
AC	22.00	23.60	34.10	25.20	33.55
SD	4.80	8.35	25.92	11.98	31.55
AV	692.2	754.8	1243.3	797.4	1049.9
SD	377.6	444.4	1022.4	584.1	1089.2
AT	50.3	78.3	718.1	138.6	1384.7
SD	46.8	65.7	1999.4	217.7	3803.1
AS	5.51	5.69	4.47	5.46	4.66
SD	0.94	1.47	1.60	1.56	1.92

We have reached to the same conclusions by comparing the performance of the two methods for $q = 3$. However, the growth of AV as a function of q is much faster in the method derived in [15], where objective functions with $q > 3$ would require more efficient procedures for solving the problem in the outcome space. On the other hand, as the growth of AV is *delayed* by Algorithm A_2, comparatively larger problems can be solved. Table 4 presents the results obtained by Algorithm A_2 for $q = 5$. The average number of times that Algorithm A_1 is invoked by Algorithm A_1 (AS) increases very slowly.

As a final remark, it is worth mentioning that as long as the objective function is a product of linear and quadratic (convex) functions, problem (P_W) will be a convex quadratic programming problem, for which very efficient solvers are available.

5   Conclusions

An algorithm for convex multiplicative problems inspired in the generalized Benders decomposition has been proposed in this paper. Connections between convex multiobjective and multiplicative programming based on existing results from the multiobjective programming literature have been established. In particular, some properties related to the concept of efficient solution have been used to derive progressively better outer approximations of the multiplicative problem. Convex duality theory has been employed to decompose the multiplicative problem into a master, quasiconcave subproblem in the outcome space, solved by vertex enumeration, and a min-max subproblem, coordinated by the master subproblem.

Numerical experience has shown that the computational effort invested in generating deepest cuts in the outcome space through the solution of min-max subproblems is compensated by a faster convergence of Algorithm A_2. The use of deepest cuts also limits the growth of vertices in the master subproblem and enables its effective solution by vertex enumeration. An adjacency list algorithm that takes into account that only efficient vertices need to be evaluated has been implemented.

The overall algorithm is easily programmed by using standard optimization packages. Further properties of the algorithm as well as its extension to more general multiplicative and fractional global optimization problems are under current investigation.

References

1. Konno, H. and T. Kuno, Multiplicative programming problems, In R. Horst and P. M. Pardalos (eds.), *Handbook of Global Optimization*, pp. 369-405, Kluwer Academic Publishers, Netherlands, 1995.
2. Benson, H. P. and G. M. Boger, Multiplicative programming problems: Analysis and efficient point search heuristic, *Journal of Optimization Theory and Applications*, 94, pp. 487-510, 1997.
3. Benson, H. P., An outcome space branch and bound-outer approximation algorithm for convex multiplicative programming, *Journal of Global Optimization*, 15, pp. 315-342, 1999.
4. Geoffrion, A. M., Elements of large-scale mathematical programming, *Management Science*, 16, pp. 652-691, 1970.
5. Floudas, C. A. and V. Visweswaram, A primal-relaxed dual global optimization approach, *Journal of Optimization Theory and Applications*, 78, pp. 187-225, 1993.
6. Thoai, N. V., Convergence and application of a decomposition method using duality bounds for nonconvex global optimization, *Journal of Optimization Theory and Applications*, 133, pp. 165-193, 2002.
7. Geoffrion, A. M., Generalized Benders decomposition, *Journal of Optimization Theory and Applications*, 10, pp. 237-260, 1972.
8. Yu, P-L., *Multiple-Criteria Decision Making*, Plenum Press, New York, 1985.
9. Horst, R., P. M. Pardalos and N. V. Thoai, *Introduction to Global Optimization*, Kluwer Academic Publishers, Netherlands, 1995.
10. Geoffrion, A. M., Solving bicriterion mathematical programs, *Operations Research*, 15, pp. 39-54, 1967.
11. Katoh, N. and T. Ibaraki, A parametric characterization and an ϵ-approximation scheme for the minimization of a quasiconcave program, *Discrete Applied Mathematics*, 17, pp. 39-66, 1987.
12. Lasdon, L. S., *Optimization Theory for Large Systems*, MacMillan Publishing Co., New York, 1970.
13. Chen, P. C., P. Hansen and B. Jaumard, On-line and off-line vertex enumeration by adjacency lists, *Operations Rsearch Letters*, 10, pp. 403-409, 1991.
14. MATLAB User's Guide, The MathWorks Inc., http://www.mathworks.com/
15. Kuno, T., Y. Yajima and H. Konno, An outer approximation method for minimizing the product of several convex functions on a convex set, *Journal of Global Optimization*, 3, pp. 325–335, 1993.

High-Fidelity Models in Global Optimization

Daniele Peri[1] and Emilio F. Campana[2]

[1] INSEAN, Via di Vallerano, 139 - 00128 - Roma, Italy
D.Peri@insean.it,
http://rios4.insean.it
[2] INSEAN, Via di Vallerano, 139 - 00128 - Roma, Italy
E.Campana@insean.it

Abstract. This work presents a Simulation Based Design environment based on a Global Optimization (GO) algorithm for the solution of optimum design problems. The procedure, illustrated in the framework of a multiobjective ship design optimization problem, make use of high-fidelity, CPU time expensive computational models, including a free surface capturing RANSE solver. The use of GO prevents the optimizer to be trapped into local minima.

The optimization is composed by global and local phases. In the global stage of the search, a few computationally expensive simulations are needed for creating surrogate models (metamodels) of the objective functions. Tentative design, created to explore the design variable space are evaluated with these inexpensive analytical approximations. The more promising designs are clustered, then locally minimized and eventually verified with high-fidelity simulations. New exact values are used to improve the metamodels and repeated cycles of the algorithm are performed. A Decision Maker strategy is finally adopted to select the more promising design.

1 Introduction

Simulation-Based Design (SBD) in the engineering design community context still suffers from some major limitations: first, while real design problems are multiobjective, practical applications are mostly confined to single objective function problems; second, it is relying exclusively on local optimizers, typically gradient-based, either with adjoint formulations or finite-differences approaches; third, the use of high-fidelity, CPU time expensive solvers is still limited by the large computational effort needed in the optimization cycles so that simplified tools are still often adopted to guide the optimization process.

The availability of fast computing platforms and the development of new and efficient analysis algorithms is partially alleviating the third limitation. However, when the evaluation of the objective function involves the numerical solution of a partial differential equations (PDE, such as the Navier-Stokes equations to give a real-life example) a single evaluation might take many hours on current generation of computers. For this reason the number of PDE-constrained optimization in ship design are still limited: [13], [30], [22], [18]. Moreover, the use of

C. Jermann et al. (Eds.): COCOS 2003, LNCS 3478, pp. 112–126, 2005.

these computational expensive models is still confined to single objective problems solved with local optimizers. Recent advances ([23], [25], [24], [17]) were dedicated to address this challenge, expanding optimization applications from single- to multiobjective problems.

Our present goal is the development of a new optimization procedure to solve multiobjective problems searching for the global optimum, overcoming the aforementioned limitations through the use of metamodels and of an alternate global-local stage in the algorithm. The intent of the present paper is to illustrate the procedure and to give numerical evidence of its capability.

2   Open Problems in SBD

When dealing with a complex ship design, open problems in SBD are easily recognized. Optimization typically involves a large number of design variables and a number of different disciplines and objectives, requiring hundreds or thousands of function evaluations to converge to an optimal design. It is also clear that (1) the size of the design space increases exponentially with the number of design variables, (2) both gradient-based optimization methods, which need the evaluation of the gradient components of the objective functions, and gradient-free pattern search methods become more and more expensive with the use of complex, high-fidelity CFD solvers as analysis tools and (3) nonconvex feasible design variable space and multimodal objective functions (i.e. functions with many local minima) can trap local optimizers in local minima preventing these from locating the best design, while the use of Global Optimization (GO) algorithms would lead back to a further increase of the computational efforts.

While item (1) is obviously an unavoidable consequence of the complexity of the design problem, a number of possible strategies to face item (2) exist. When using gradient-based optimization methods, the control theory approach allows for dramatic computational cost advantages over the finite-difference method of calculating gradients, being substantially independent of the number of design variables. Sensitivity Equations Methods and Adjoint Methods belong to this class. However, even if automatic differencing compilers exists (for example TAF, Transformation of Algorithms in Fortran [8]) these approaches often require an "appropriate code preparation" [31], i.e. a development phase on the source code of existing in-house CFD solvers, which is not always negligible and openly recognized.

The second alternative strategy is to utilize global approximation models which are often referred to as metamodels, as they provide a "model of the model", partially replacing expensive simulation models during the design and optimization process. Metamodelling techniques have been widely used for design evaluation and optimization in many engineering applications (for reviews of metamodelling applications in structural and multidisciplinary optimization see [1] [27]).

With regard to item (3), another key issue in ship design optimization will become the use of GO methods [32]. Many engineering applications use accepted methods for single-extremum function minimization without any prior investigation of unimodality. Researchers of the ship design community often recognize the problem of starting from a "good" parent hull form but the consideration of rigorous practical GO procedures have been outside their attention. However, as recalled in item (3), often the feasible design spaces are nonconvex and the objective functions distribution in the design variable spaces are multimodal, so that even a simplified design problem includes many local optima that can trap local optimizers. In such design variable spaces, unsophisticated use of local optimization techniques is normally inefficient (an example will be given in the next paragraph.) GO algorithms are hence important from a practical point of view and should be used despite an increase of the computational effort.

In order to handle such difficulties the considered design problem is converted into a GO one. The solution strategy of GO methods consists of a global stage and local stages. A uniform covering method, namely the LP_τ grid of Sobol [29] is adopted to get global information on the objective functions and to explore the design variable space.

Design Of Experiment (DOE) is chosen for the initial construction of the metamodels: from now on, the high-fidelity solvers are only applied for verification once a promising solution is detected by the actual metamodels. Successive high-fidelity computations (used for verifying the promising solutions) are added to DOE to enlarge the training set of the metamodels, increasing also their reliability.

In the present paper, the GO procedure has been used to solve a multiobjective problem for the DDG51. Five objective functions have been considered, governed by three different PDEs; again, the RANSE code has been used for the prediction of the free-surface flow past the ship.

3   Why Global Optimization?

Before we start the description of the developed GO procedure, an example is given to illustrate how local optimization is rarely the appropriate technique for shape optimization, even in a simple design variable space. Indeed, due to nonlinear constraints (even a simple equality constraint on the displacement is non linear), nonconvex feasible design spaces are quite common in practical problems (e.g. [16]) as well as multimodality of the objective functions, and local optimization techniques are inefficient in solving these problems. A simple ship design problem for the S175 Containership is presented. The goal is the minimization of the peak value of the Response Amplitude Operator (RAO, a measure of the ship's response in waves analyzed in the frequency domain) for the heave motion in head seas, for the ship advancing at 16 knots. Only six design variables are used in this simple numerical test. As geometrical constraints, lower and upper bounds on the beam and on the ship's displacements are enforced. A simple strip-theory code is used as analysis tool.

The inadequacy of the local optimization approach may be observed in Figs. 1 and 2. In Fig. 1 the RAO for the original hull is reported versus those of

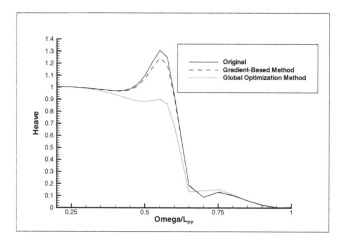

Fig. 1. Optimization results on the S175 containership as from different optimization algorithms: solid line is the original hull shape RAO (heave motion in head seas, speed of 16 kts), dashed line represents the result obtained by means of a standard gradient-based method, dotted line is the result of a Global Optimization Method

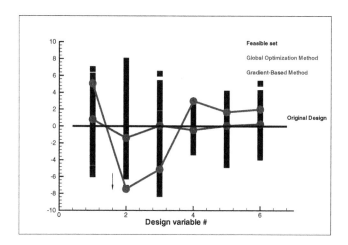

Fig. 2. Optimization results on the S175 containership as from different optimization algorithms in the design variable space. Each column represents a design variable. Holes in the columns indicate a non-connected region in the feasible space. Blue line represents the result obtained by means of a standard gradient-based method, red line is the result of a Global Optimization Method

two optimized solutions. The dashed line indicates the design obtained with a standard gradient-based local optimization technique while the dotted line represents the performances of the globally optimized design.

The local optimization procedure is able to improve the original design obtaining a new shape which displays a reduction of the RAO's peak value. However the reduction is small and a much better result might be obtained using the GO approach.

The result can be easily understood when looking at Fig. 2, where the histograms of the design variables distribution have been reported for all the designs of the feasible solution set constructed by the GO procedure. The "holes" in the distribution of the feasible solutions (gaps in the design variables No. 2,3 and 6) are caused by the nonlinear constrains.

In the histogram the original ship is given by assuming all the design variables equal to zero. Blue symbols indicate values of the solution obtained with the local optimization while red ones are those of the GO technique. It is clear that, starting from the original shape, the local optimization algorithm would never be able to jump across the holes created by the nonlinear constraints and that the local optimizer has been trapped by a local optima which was very close to the initial design. On the contrary, the GO algorithm explores the whole design variable space without remaining trapped in the local minima or confined by the nonconvex design space.

4   Description of the GO Algorithm

This section is devoted to the description of the developed algorithm and its definition in the group of GO techniques. For an extensive coverage of various methods of GO, useful reference is [32].

The algorithm is illustrated in the case of a multiobjective problem but it is also applicable to solve single objective function problems. Different ways of classifying Global Optimization methods exist. The proposed method belongs to the class of deterministic optimizer, and to the family of Covering Methods, but with some features similar to the Adaptive Clustering COvering (ACCO) - Adaptive Clustering covering with Descent (ACD) schemes proposed in [28]. It is basically founded on the consideration that the only way to find out the global minimum of an unknown objective function, whose global characteristics of continuity are not available, is to search uniformly the design parameter space.

The algorithm hence consists of two main stages: (i) a global search phase, where a GO algorithm is used to explore the design space avoiding local minima and trying to locate regions where attractive solutions are found and (ii) a local refinement phase, where best configurations (according to the Decision Maker) are grouped in clusters and then locally optimized with a multiobjective gradient-based technique. The fundamental elements of the global algorithm are described in the following.

Formulation of the GO problem: In the most general way, the GO problem for a single objective function can be stated as follows. Let us consider a function $f : Z \rightarrow R$, where $Z \subset \mathcal{R}^N$:

$$\text{find } \mathbf{x} \in Z$$
$$\text{such that}$$
$$f(\mathbf{x}) = min f(\mathbf{x}), \ \mathbf{x} \in Z.$$

In the constrained problem, bounds on the N design variables d and m functional constraints should be considered:

$$g_j(\mathbf{x}) \leq 0, \ j = 1, ..., m$$
$$d_j^L(\mathbf{x}) \leq d \leq d_j^U(\mathbf{x}) \ j = 1, ..., N$$

In general $f(\mathbf{x})$ and $g(\mathbf{x})$ may be nonconvex, nonsmooth functions. The multiobjective optimization problem can now be easily defined:

$$\min\{ \ f_1(\mathbf{x}), \ f_2(\mathbf{x}), \ ... \ , \ f_k(\mathbf{x}) \ \}$$

where we have $k(\geq 2)$ objective functions $f_i : Z \to R$.

Manipulation of the ship geometry and of the mesh: In the implementation of an algorithm for shape optimization there is the obvious need for a geometry modification procedure and several attempts have been made to deal with this item. We decided not to rely on the use of a specific commercial CAD program but to induce modification in the ship's geometry by controlling a perturbation polynomial surface, which is added to the unmodified original geometry (details in [22]). The control points of this polynomial surface will become the design variables d of the design problem. General guidelines for this procedure are the following: (i) when only a part of the ship is directly involved by shape optimization, the modified region should join the original design without discontinuities and should be generally smooth; (ii) the number of design variables should be kept as small as possible to minimize the number of evaluations of the gradient of the objective function, but (iii) the algorithm should be as flexible as possible in order to achieve the most number of possible solutions.

The above requirements have been obtained by using Béziér patches gradually reducing to a zero level while approaching the unmodified hull shape. In this way, geometric continuity between grid boundaries is guaranteed and if the number of control points is kept sufficiently small, realistic geometry can be obtained that do not need major refinements prior to construction. Once the geometry is modified, the volume grid is adjusted accordingly.

Metamodel identification from CFD analysis results: As recalled before, when the number of the design variables increases calculations of the gradient components become more expensive. Under this perspective, the application of CFD for the evaluation of the objective functions is discouraged. Anyway, the possible existence of a (unknown) relationship between the results coming from CFD and a much simpler analysis tool could help in reducing the number of solutions addressed to the high-fidelity model. An interesting possibility is to explore the analytical structure of the data coming from the CFD over the design variables space trying to derive a metamodel. Obviously, a specific surface must be constructed for each different objective function considered in the multiobjective problem.

A variety of metamodelling techniques exist (for an analysis of their performances see [15]). Polynomial regression models [20] is a widely known approach for the design and analysis of computer experiments. The coefficients of the polynomial functions may be computed by using a least square technique, or a more sophisticated identification parameter technique, as the Levemberg-Marquard method. Moreover, performing the Analysis Of the VAriance (ANOVA) of the response surface it is also possible to enhance their quality, deleting terms not really affecting the approximation [9]. Obviously, this method is capable to correctly describe only objective functions up to a certain order. If the objective function is more complicated, the degree of the polynomial may be increased, but the metamodel quality may decrease because of numerical instabilities.

Artificial neural networks [26], [5] are well-known approaches for identifying approximations of complex simulation codes and to fit a wide class of objective functions. In the following, a neural network of radial basis function (RBF neural network) has been selected as metamodel to solve the GO problem. Given a set of T points, the interpolating function has the form:

$$y(\mathbf{x}) = \sum_{t=1}^{T} w_t \Phi(||\mathbf{x} - \mathbf{x^t}||)$$

where Φ is a continuous function, which can be chosen among radial functions. The most used function in an RBF network is a Gaussian

$$\Phi = e^{-r^2}$$

RBF networks are feedforward with only one hidden layer. RBF hidden layer units have a receptive field, which has a centre (a particular input value at which they have a maximal output). Their output tails off as the input moves away from this point.

To construct the metamodel one has to select the T training points that have to be computed with the high-fidelity model. This can be performed using a Design Of Experiment (DOE) technique. A complete factorial design usually requires at least L^N solutions to be computed with the high-fidelity model, where N is the number of design variables and L is the number of levels in which each design variable interval is subdivided. The minimum value is 2N, the vertices of a hypercube constructed in the design variables space around the initial design. Consequently, the number of solutions needed to build the metamodel with a complete factorial design rapidly grows. Hence, an incomplete factorial design, in which some extremes of the design variable space are discharged, is usually applied. The criterion for the vertices elimination generates a huge number of different methodologies. In this paper we have selected an Orthogonal Array (OA) technique [12], for which only N+1 points (i.e. CFD solutions) are requested to build the metamodel.

The search in the design variables space: Once an interval for the design variables has been fixed, trial points (i.e. solutions to be evaluated with the metamodel) must be distributed into the design space. Uniformity of the distribution of the trial points is a crucial characteristic for the success of the optimization, since an under-sampling of some region could deceive the optimizer forcing to discharge that portion of the design space. A regular sampling (cubic grid) would produce an uniform hexahedrical mesh on the hypercube defined in the design space, with two major drawbacks: too many points are needed when the number of design variables increases, and a marked shadow effect is produced (i.e. the coincidence of the projections of some points on the coordinate axes). LP_τ grids [29], belonging to the family of the Uniformly Distributed Sequences (UDS), have some attractive features, like an high degree of uniformity with a reduced set of trial points and a moderate shadow effect. In [29] the maximum number of points in the LP_τ distribution is 2^{16} and this value has been selected to sample the design space during the numerical test.

Once an UDS is placed in the design variables space, geometrical constraints are verified on these configurations and a discrete approximation of the Feasible Solution Set (F_{SS}) is obtained. The density of trial points belonging to the F_{SS} is clearly connected with the adopted UDS and it is desirable the number of points in the F_{SS} to be as highest as possible. On the other hand, CPU time needed for the constraints verification may be not negligible for real applications and a very long time for constructing the F_{SS} may be necessary in the case of an excessive sampling. Anyway, local refinement techniques described below aid to reduce this problem, allowing for an increase of the number of points in the F_{SS} near the best configurations during the process of optimization.

Pareto optimality: In multiobjective problems, the problem of finding optimal solutions among those belonging to the feasible set is solved employing the concept of Pareto optimality:

Definition: a configuration identified by the objective vector $\mathbf{x_o}$ is called *optimal in the Pareto sense* if there does not exist another design $\mathbf{x} \in F_{SS}$ such that $f_k(\mathbf{x}) \leq f_k(\mathbf{x_o}) \forall k = 1, ..., K$, where K is the number of objective functions, and $f_k(\mathbf{x}) < f_k(\mathbf{x_o})$ for at least for one $k \in [1, ..., K]$.

By applying the Pareto definition on the feasible solutions it is possible to find all the design vectors where none of the components can be improved without deterioration of one of the other components (non-dominated solutions). These designs belong to the Pareto optimal set $\mathcal{P}$.

Decision maker: Mathematically, all $\mathbf{x} \in \mathcal{P}$ are acceptable solutions of the multiobjective problem. However, the final task is to order the design vectors belonging to $\mathcal{P}$ according to some preference rules indicated by the designer and select one optimal configuration among them. In general, one needs the co-operation between the decision maker and the analyst. A decision maker may be defined [19] as the designer who is supposed to have better insight into the problem and can express preference relations between different solutions. On the contrary the analyst can be a computer program that gives the information to the decision maker. A wide number of different methodologies exist in literature (see [19] for an extensive summary of the subject) depending on the role of the decision maker in the optimization process. In no-preference methods the opinions of the decision maker are not taken into consideration and the selection is accomplished by measuring (in the objective function space) the distance between some reference points (the "ideal" objective vector) and the Pareto solutions. It is the simple Global criterion, which can use different metrics. A powerful and classical way to solve this problem from the standpoint of practical application is the use of the ideas of goal programming. The procedure is simple: a list of hypothetical alternatives with assigned values (aspiration levels) of the objective functions is ranked by the designer according to his preference and experience (these are the goals). The problem is then modified into the minimization of the distance from these goals. Designer data may be used for constructing a metamodel of the preference order. In this way, the optimization process will be mainly driven by the real needs of the designer, and the portion of Pareto set explored by the optimizer will contain the subset of the most preferable solution in the opinion of the designer.

Local refinement of the best solution: During the development of the optimization process some designs show better characteristics than others, and the probability that the optimal solution is located in the vicinity of these points is high. As a consequence, those portions of F_{SS} around promising points are deemed more interesting than others. The local refinement may follows two different strategies: (i) use a local method, able to give small improvements for all the objective functions in the neighborhood of a promising point, and/or (ii) adopt a clustering technique [3] in order to identify the region for which a deeper investigation is required and an increased density of trial points is wanted.

Cluster analysis of the promising solutions and refinement of the clusters: The task of any clustering procedure is the recognition of the regions of attraction, i.e. those regions of the objective space such that for any starting point x, an infinitely small step steepest descent method will converge on an essential global minimum [32]. In a multiobjective formulation the points belonging to the Pareto optimal set are the most promising solutions for the problem under consideration. Hence, we decide to assume these points as centers of regions of attraction. Some clustering algorithms are described in [32]. Here, the one proposed by Becker and Lago [3] has been adopted.

The local refinement of the F_{ss} is then obtained by placing a reduced LP_τ net, with smaller radius and fewer points, around the center of the clusters. The radius of the investigated region in the neighborhood of the Pareto point decreases during the optimization process. The distribution is rotated at each step, in order to spread out points in all the directions.

The algorithm for the GO problem: Main steps of the algorithm may be summarized as follows:

1. Initial exploration of the design space - Orthogonal Array is adopted for the initial exploration of the design space and trial points are distributed.
2. Model Identification from CFD results - Trial points are evaluated using the CFD for the construction of the metamodels (one for each objective function);

3. The search in the design variables space - New trial designs (2^{16} points = 65536) are uniformly distributed in the design variable space by using the LP_τ-grid;
4. Derive the feasible set - Enforcing the geometrical and functional constraints (i. e. stability) a large part of these trial points is discharged and the feasible solution set is derived;
5. Identify the Pareto front - Analyse feasible points using the metamodels and find all $\mathbf{x} \in \mathcal{P}$;
6. Adopt a Decision Maker strategy for ordering the designs and finding non-dominated solutions;
7. Local refinement of the best solution with a multiobjective gradient step based on the metamodels (or with a scalarization of the problem according to the DM);
8. Verification of the best solution using the CFD solvers. The new solution will be added to the metamodel for its improvement;
9. Clustering of the Pareto solutions is performed in the design space around dominating solutions identified by the DM. A reduced number of sets is obtained;
10. Refinement around the center of the clusters: new trial designs are uniformly distributed with smaller LP_τ-grids centered around the clusters;
11. go to step 4 until no more regions of attraction are found;

An important feature of this algorithm is that, as a consequence of both the refinements (steps 7 and 10), new added points may fall in a region of the design space which was not considered in the initial distribution of trial points. This is a useful quality of the method: in fact, in our particular case, since there is not a strong connection between design variables (the control points of the Béziér patches) and geometrical constraints, a correct estimation of the boundaries of the design parameters is non trivial. For this reason, the initial distribution does not cover the whole F_{SS}, since the investigated volume must be as small as possible in order to retain the point density of the F_{SS}. The local refinement technique automatically corrects the underestimation of the design space extension: the optimization problem is still constrained, but bounds on the design variables may change dynamically within the course of the optimization problem solution.

5   Multi-objective Optimization Test

A multiobjective problem for a frigate ship (model 5415 of the David Taylor Model Basin, an early design of the DDG51 of the US Navy) has been set up as described in the following.

Objective functions description: The goal is the minimization of five objective functions at the service speed (20 knots):

- Function F_1 is the wave resistance of the ship at the service speed, as computed by using a non-linear panel solver for steady free surface potential flow [2].
- Functions F_2 and F_3 represent some seakeeping performances of relevance in the definition of the operability of the ship: they are respectively the peak values of the heave and the pitch motions in head seas. Their values are estimated by applying a 3D panel code in the frequency domain (the FreDOM code, details in [14].
- To compute functions F_4 and F_5, the MGShip RANSE solver for steady free surface flows [6] has been used. F_4 represents a measure of the uniformity of the axial velocity at the propeller disk, considered a relevant parameter in the design of the ship's propeller, while F_5 is related to the minimization of the vortices produced at the junction of the sonar dome with the bare hull, expressed as the mean of the vorticity in an area just behind the dome. F_4 and F_5 control regions are two circles or radius 0.018 L_{PP} and 0.014 L_{PP} respectively, placed at X=7.1m and X=113.6m from the fore perpendicular. Control region of F_5 has been positioned on the base of the location of the sonar dome vortices, as seen from experimental measurements.

It may be observed that the wave resistance of the DDG51 could have been computed by using the RANSE free surface code by applying on the numerical wave pattern some linear method for extracting the wave resistance information (longitudinal or transverse cut methods). However, we have had the feeling that the free surface grid used in the RANSE computation inside the optimization process was not enough dense to capture correctly the wave pattern. For this reason the nonlinear panel code has been preferred for the evaluation of F_1. Moreover, a test about the connection between different solvers in a multidisciplinary framework was of great interest in the construction of the optimizer and to add an additional solver was helpful under this perspective.

Design variables definition: For the optimization of the hull shape, 15 design variables have been used, acting both on the side of the entire hull and on the bulb. Stem and transom stern have been left unchanged, as well as the ship length L_{PP}. The ship modification is performed by means of superposition of three different Béziér patches to the original ship geometry, two acting in the y direction, for the hull and the bulb, and one acting in the z direction for the bulb geometry only.

Geometric and functional constraints: A specific (nonlinear) constraint is applied on the total displacement: a maximum variation of about (2% is allowed. Bounds on the design variables are also enforced, even if those limits are trespassed during the cycles, as explained before.

Numerical solvers and conditions for the test: Summarizing, three different solvers are applied: a non-linear potential flow solver (for the evaluation of F_1), a potential solver in the frequency domain for the prediction of the response of the ship in waves (F_2 and F_3) and a surface capturing RANSE steady solver (F_4 and F_5). Consequently, five metamodels are constructed using RBF neural networks.

All the computations but the seakeeping ones are performed for the ship free to sink and trim, and control regions for F_4 and F_5 are translated consequently.

Numerical results: An Orthogonal Array (OA) of 16 elements has been applied for the DOE phase for the metamodel, obtained by adopting a RBF neural network. During the initial search in the design space 65536 points have been disseminated. After having enforced the constraints, about 11064 designs fall inside the F_{SS}. After 140 optimization cycles (which imply 157 objective function evaluations with the high-fidelity solvers: $140 + 16$ for the initial construction of the metamodels with the OA $+ 1$ for the original design), 15273 points belong to F_{SS}, due to the effect of the clustering and resampling, with a mean of about 30 new points added to F_{SS} per each iteration. While the optimization is proceeding, the number of the new design solutions added to the F_{SS} per cycle is increasing, because the algorithm focuses the resampling area in progressively smaller regions near the Pareto solutions.

The final resulting Pareto front, restricted only to the configurations showing a reduction of all the objective functions, is reported in Table 1. Results are reported in a table because of the impossibility of representing the Pareto front in $\mathcal{R}^N$. Each column reports an objective function, non-dimensionalized by its initial (original) value. First column indicates an identification number of the configuration, and the last two columns report the mean value and variance of the objective functions, giving an indication about the homogeneity of the enhancements. All the designs for which one objective function displays the best performance are reported in bold.

Table 1. Pareto Optimal Set for the here depicted test case. All the objective functions are non-dimensionalized by their initial value: configurations showing values greater than unit have been deleted. Best values for each objective function are plotted in bold

ID. #	F_1	F_2	F_3	F_4	F_5	Mean F_i	σ
39	0.58279	0.94629	0.97535	0.76555	**0.22946**	0.69988	0.27420
41	0.54586	0.97397	0.98408	0.72584	0.29821	0.70559	0.26133
53	0.57674	0.94155	0.97050	**0.61619**	0.51789	0.72457	0.19177
57	**0.52733**	0.96629	0.98395	0.66259	0.53376	0.73478	0.20216
59	0.58851	**0.89009**	**0.90445**	0.75741	0.58808	0.74570	0.13838
29	0.57447	0.98353	0.98154	0.75522	0.48378	0.75570	0.20479
35	0.55503	0.97043	0.97903	0.77654	0.77882	0.81197	0.15582
21	0.84752	0.92517	0.95868	0.70223	0.64447	0.81561	0.12300
48	0.72060	0.93431	0.96585	0.68400	0.77633	0.81621	0.11363
67	0.57716	0.89578	0.93559	0.99313	0.75175	0.83068	0.14976
89	0.66554	0.85228	0.91643	0.94391	0.89387	0.85440	0.09909
69	0.99273	0.93034	0.95144	0.98579	0.54357	0.88077	0.17013

Data are ordered by the mean value. Configuration #59 is an interesting solution, being the one with the lowest variance among those designs displaying enhancements w.r.t. all F_i. Some of the extreme designs may show significant worsening on one F_i and of course they do not represent interesting solutions for the optimization problem. However, they give an idea about the extremes of the boundary of the Pareto set. Also, differences in the hull-forms are relevant, giving account of the volume of the investigated design space.

Configuration #57 is the best for resistance (function F_1). Design #59 is the best as to the heave and pitch response.

Configuration #53 is the best for the uniformity of the flow at the propeller plane Obtained results are very encouraging, being the mean value of the axial component of the velocity at the propeller disk greatly enhanced, and also its variance being reduced.

Finally, configuration #39 is the best for the flow quality behind the sonar dome. The core of the dome vortex is smoothed out, and the primary objective is obtained: in fact, the non-dimensional value of the objective function is of the order of 0.2, with a reduction of about 80%. The reduction of the vorticity will result in a reduction on the flow noise in this region. This is a promising result for the design of ship's sonar dome which can be very difficult to be obtained by using traditional design approaches, guided only from the experience of the designers. Indeed, the identification of the those hull parameters affecting the vorticity production is not easy.

As a final comment, when more than a single design criteria is assumed (as is always in the real design), the task is for the designer becomes complex, and there is no guarantee that a good solution can be found by traditional design process.

6   Conclusions

Optimisation tools could help the designer and GO techniques can lead to new design concepts. The final goal of the authors is to develop a useful GO solver for ship design and techniques for reducing CPU-time needs, a fundamental step if a GO problem has to be solved. To this aim, a GO problem in a multiobjective context has been formulated and solved with an original algorithm. Although the numerical results are still preliminary, strong reductions on the interesting quantities have been obtained. The applied numerical solvers are able to give reliable information on the flow field, allowing improvements otherwise difficult to be obtained in the absence of correlations law between main geometrical parameters and local flow variables. The optimization tool seems to be able to co-ordinate the different objectives and the analysis tools used in the procedure are used in a rational way. The inclusion of this approach into the spiral design cycle is recommended, in particular when some special requests are present in the design specifications.

Acknowledgements

This work has been supported by the U.S. *Office of Naval Research* under the grant No. 000140210489, through Dr. Pat Purtell. The Authors also wish to thank Natalia Alexandrova for her technical advices, Andrea Di Mascio for the use of MGShip RANSE solver and Andrea Colagrossi for the use of the FreDOM seakeeping solver

References

1. Barthelemy, J.-F.M.; Haftka, R.T.: Approximation concepts for optimum structural design - a review. Structural Optim. 5, (1993) 129-144
2. Bassanini P., Bulgarelli U., Campana E.F., Lalli F.: The wave resistance problem in a boundary integral formulation. Surv Math Ind, 1994-4.
3. Becker R.W., Lago G.V.: A global optimization algorithm Proceedings of the 8th Allerton Conference on Circuits and Systems Theory, (1970) 3-12.
4. Chang, K.J., Haftka, R.T., Giles, G.L. and Kao, P.-J.: Sensitivity-based scaling for approximating structural response. Journal of Aircraft, 30(2) (1993), pp.283-288.
5. Cheng, B., Titterington, D.M.: Neural networks: a review from a statistical perspective. Statistical Sci. 9i (1994), 2-54
6. Di Mascio A., Broglia R., Favini B.: A second-order Godunov-type scheme for Naval Hydrodynamics. Godunov (Ed) Methods: Theory and application, Singapore (2000): Kluwer Academic/Plenum.
7. Dixon L.C.W., Szegö G.P.: Towards global optimization, (1975) North-Holland.
8. Giering, R., and Kamnski, T.: Recipes for Adjoint Code Construction. ACM Trans. on Math. Software, Vol. 24 (1998), No. 4;437-474.
9. Giunta A.A., Balabanov V., Kaufman M., Burgee S., Grossman B., Haftka R.T., Mason W.H., Watson L.T.: Variable-Complexity Response Surface Design of an HSCT configuration. Multidisciplinary Design Optimization, Alexandrov N.M. and Hussaini M.Y. eds (1997), SIAM, Philadelphia, USA.
10. Haftka R.T., Vitali R., Sankar B.: Optimization of Composite Structures Using Response Surface Approximations. NATO ASI meeting on Mechanics of Composite Materials and Structures, Troia, Portugal (1998).
11. Haimes Y.Y., Li D.: Hierarchical multiobjective analysis for large-scale systems: review and current status. Automatica, Vol.24 (1988), No.1, 53-69
12. Hedayat A.S., Sloane N.J.A., Stufken J.: Orthogonal Arrays: Theory and Applications. (Springer Series in Statistics) (1999), Springer-Verlag, Berlin, Germany
13. Hino, T., Kodama , Y., and Hirata , N., 1998: Hydrodynamic shape optimizationof ship hull forms using CFD. Third Osaka Colloquium on Advanced CFD Applications to Ship Flow and Hull Form Design (1998), Osaka Prefecture Univ. and Osaka Univ., Japan.
14. Iwashita, H., Nechita, M, Colagrossi, A., Landrini, M., Bertram, V.: A Critical Assessment of Potential Flow Models for Ship Seakeeping. Osaka Colloquium on Seakeeping (2000), pp. 37-64 (Osaka), Japan
15. Jin R., Chen W., Simpson T.W.: Comparative studies of metamodelling techniques under multiple modelling criteria. Struct. Multidisc. Optim. (2001), 23.
16. Knill, D.L., Giunta, A.A., Baker, C.A., Grossman, B., Mason, W.H., Haftka, R.T., Watson, L.T.: Response surface models combining linear and Euler aerodynamics for supersonic transport design. Journal of Aircraft (1999), Vol. 36, 1, pp.75-86

126 D. Peri and E.F. Campana

17. Minami,Y. , Hinatsu M.: Multi Objective Optimization of Ship Hull Form Design by Response Surface Methodology. 24th Symposium on Naval Hydrodynamics (2002), Fukuoka, JAPAN

18. Newman III, J.C., Pankajakshan, R., Whitfield, D.L., and Taylor, L.K.: Computational Design Optimization Using RANS. 24th Symposium on Naval Hydrodynamics (2002), Fukuoka, JAPAN

19. Miettinen K.M.: Nonlinear multiobjective optimization. Kluwer Academic Publisher (1999).

20. Myers R.H., Montgomery D.C.: Response Surface Methodology. Wiley, USA (1997).

21. Pareto V.: Manuale di economia politica, Società editrice libraria (1906), Milano, Italy. Also in Manual of Political Economy, The MacMillan Press Ltd (1971).

22. Peri D., Rossetti M., Campana E.F.: Design optimization of ship hulls via CFD techniques. Journal of Ship Research, 45 (2001), 2, 140-149.

23. Peri D., Campana E.F., Di Mascio A.: Development of CFD-based design optimization architecture. 1st MIT conference on fluid and solid mechanics (2001), Cambridge, MA, USA.

24. Peri D., Campana E.F.: High fidelity models in the Multi-disciplinary Optimization of a frigate ship. 2nd MIT conference on fluid and solid mechanics (2003), Cambridge, MA, USA.

25. Peri D., Campana E.F.: Multidisciplinary Design Optimization of a Naval Surface Combatant Journal of Ship Research, 47 (2003), 1, 1-12.

26. Smith M.: Neural networks for statistical modeling. Von Nostrand Reinhold (1993), New York.

27. Sobieszczanski-Sobieski, J., Haftka, R.T.: Multidisciplinary aerospace design optimisation: survey of recent developments. Struct. Optim. 14 (1997),1-23

28. Solomatine D.P.: Two strategies of adaptive cluster covering with descent and their comparison to other algorithms. Journal of Global Optimization (1999), 14, 1, 55-78.

29. Statnikov R.B., Matusov J.B.: Multicriteria optimization and engineering. Chapman & Hall (1995), USA.

30. Tahara, Y., Patterson, E., Stern, F., and Himeno, Y.: Flow- and wave-field optimization of surface combatants using CFD-based optimization methods. 23rd ONR Symposium on Naval Hydrodynamics (2000), Val de Reuil, France.

31. Thomas J.P., Hall K.C., Dowell E.H.: A discrete adjoint approach for modeling unsteady aerodynamic design sensitivities. 41-th Aerospace Science Meeting and Exibit (2003), Reno, Nevada, USA

32. Törn A.A., Žilinskas A.: Global optimization. Springer-Verlag (1989), Berlin, Germany

Incremental Construction of the Robot's Environmental Map Using Interval Analysis

Cyril Drocourt, Laurent Delahoche, Eric Brassart,
Bruno Marhic, and Arnaud Clérentin

UPJV – IUT Département Informatique,
Avenue des facultés, 80025 AMIENS Cedex 1
{Cyril.Drocourt, Laurent.Delahoche, Eric.Brassart,
Bruno.marhic, Arnaud.Clerentin}@u-picardie.fr

Abstract. This paper deals with an original simultaneous localisation and map building paradigm (SLAM) based on the one hand on the use of an omnidirectional stereoscopic vision system and on the other hand on an interval analysis formalism for the state estimation. The first part of our study is linked to the problem of building the sensorial model. The second part is devoted to exploiting this sensorial model to localise the robot in the sense of interval analysis. The third part introduces the problem of map updating and deals with the matching problem of the stereo sensorial model with an environment map, (integrating all the previous primitive observations). The SLAM algorithm was tested on several large and structured environments and some experimental results will be presented.

1 Introduction

The stage of incremental construction of the robot's environmental map is preponderant for the increase of its autonomy [11]. It consists in managing a coherent update of the cartographic primitives' state during the robots movement. This function is directly correlated to that of the localisation : the robustness of the cartographic primitives' state estimation is linked to that of the estimation of the robot's position. In this context it is necessary to take into account the interaction between both the localisation and the modelisation errors. The interval analysis formalism provides us with an answer to this problematic. Furthermore the soundness of the localisation's paradigm and the simultaneous modelisation are tightly linked to the quantity and quality of the sensorial data. The omnidirectional vision sensor's systems are, in this case, well adapted to this constraint, especially to a stereoscopic use.

In background literature, we can distinguish two main groups of methods used to build the evolution field of a robot: the "metric" methods and the "topologic" ones.

The first approach consists of managing the notion of distance and we can find principally two types of mapping paradigm in this context :

- The first ones consist in managing the notion of distance, where the Extended Kalman Filtering (EKF) is used to build a Cartesian representation of the environment [4].

C. Jermann et al. (Eds.): COCOS 2003, LNCS 3478, pp. 127–141, 2005.

-     The second where the occupational grid notion is used to provide a metric representation. These occupancy grids manage the "occupation", the "non-occupation" or the "potential occupation" of the group of cells representing the environment. [8][1].

The second category of map representation is the topological one. This approach consists of determining and managing the location of significant places in the environment along with an order in which these places were visited by the robot. In the topological mapping step, the robot can generally observe whether or not it is at a significant place. The definition of significant places can be linked for example to the notion of "distinctive places" in the Spatial Semantic Hierarchy proposed in [14], and the notion of "meetpoints" in the use of Generalized Voronoi Graphs proposed in [3]. This kind of method is interesting to use in complement with an occupancy grid, in order to take into account the semantic aspect.

In this paper we will present an alternative method to the two main ones mentioned above. Owing to the interval analysis formalism, the presented method guarantees the environment's representation. This way, the estimation of both the robot's state and the landmarks is characterised by subpaving.

2   Sensorial Model Building

We have developed a perception system called SYCLOP, which is similar to the COPIS system used by Yagi [18]. Our system is used to achieve both the localisation and the modelisation of the environment, based on the co-operation between two sensors. The SYCLOP prototype measures 60 cm in height and is composed of a conical mirror and a CCD camera. This vision system allows us to detect vertical parts in the environment with a 2D projection onto the camera's image plane [5].

2.1   The Omnidirectional and Stereoscopic Perception System

The idea behind this co-operation is that two image acquisitions are taken at two different positions separated by a known distance d. The translation between the two positions is achieved by two horizontal rails. These rails allow us to guarantee a known rigid in-line movement between these two previous positions (Figure 1).

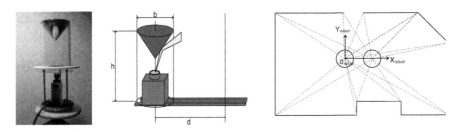

Fig. 1. Principal of the omnidirectional and stereoscopic sensor

In each acquisition, a vertical landmark of the world (doors, corners, edges, …) is characterised onto the image plane by a strongly contrasting radial straight line.

If the same radial straight line is matched in both conical images, it is quite simple to compute the location of the intersection point in the robot's reference frame. This point corresponds to a vertical landmark. This can be extended to all pairs of matched radial straight lines (Figure 1).

It is necessary to specify that the calibration of the vision system has been done before applying any sort of image processing.

The reader can find further information about the complete calibration of the SYCLOP sensor in [2](Cauchois *et al*, 1999).

2.2 The Sensorial Primitives Calculation

Our goal is to match the angular sectors of homogenous grey levels in the two images. These sectors are delimited by the radial straight lines mentioned above.

All the radial straight lines in a conical image converge to a single point called O (the projection of the revolution axis of the cone onto the image plane). This means that only the angular reference determines a radial line in the image. Thus a 2D image processing can easily be reduced to a 1D computation.

We therefore consider a concentric circle of a grey level on the image, centred on the previous point O. In order to obtain a maximal density of 1D signal information, this circle is designed on the periphery of the conical image. A 1D grey level signal is computed to characterise each image.

We have applied a segmentation algorithm based on a derivative filtering of the 1D grey level signal in order to proceed to the matching step. The reader can find more details on this method in [7]. In our case, the matching phase consists in matching two by two all the detected grey level sectors of the two stereoscopic images. As the robustness of the matching is primordial, we will use several different complementary criteria. The criteria will be merged according to the Dempster-Shafer combination rules.

As the viewpoint is different for the two images (shifted by the distance d), the landmarks in both images cannot be observed in the same way. We have retained four significant and robust criteria :

- The inclination of the approximate lines corresponding to the set of sector grey level,
- The average of the grey level of the sector

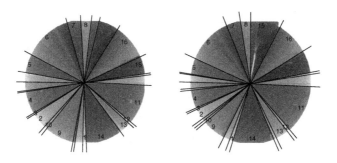

Fig. 2. Segmentation and final matching of sectors for an acquisition

- The standard deviation of the grey level of the sector.
- The geometric constraints of the sector imposed by the view point ; which can be categorised as a "simplified epipolar geometry".

We use the Dempster Shafer theory to perform the fusion [6][16]. The Dempster-Shafer method also enables us to function with partial knowledge. A final example of matching is given in Figure 2, where we can see that a large number of sectors are correctly matched [7].

Once the mutual matching of sectors has been achieved, all that we need to do is to calculate the co-ordinates of the segment points that they represent. We know the orientation angle of the two straight lines that border the sides of the angular sectors and the distance d that separates the two cones (the two images). The co-ordinates of all the points in the sensor's reference frame (situated on the centre of cone **O**) are calculated through triangulation using the following formulas :

$$x = \frac{d \times \tan(\beta)}{\tan(\beta) - \tan(\alpha)} \qquad y = \frac{d \times \tan(\beta) \times \tan(\alpha)}{\tan(\beta) - \tan(\alpha)} \qquad (1)$$

3 Localisation of a Mobile Robot Using Interval Analysis

When the imprecision is not taken into account, the localisation / modelisation process is rendered incomplete, and therefore the influence of the error of the robot's position estimation on the estimation of the vertical landmarks' parameters cannot be processed, whilst this is a main factor. There actually is an obvious interaction between the committed errors with regards to the robot's position and those introduced by the calculation of the position of the landmarks. It is this interaction that – in the process of incremental construction – is at the origin of the cumulative errors. This is the reason why we wish to present an alternative that allows to integrate the imprecision notion as of the stage of localisation and therefore, we decided to use interval analysis method.

3.1 Localisation of a Mobile Robot Using SIVIA

The SIVIA (*Set Inversion Via Interval Analysis*) algorithm was developed by Luc Jaulin and Eric Walter [12]. It enables us to determine the solution of the set inversion problem *via* subpaving (rectangular-sub-sets). The subpaving gives an approximate but guaranteed solution.

The algorithm consists in sub-dividing an initial box into two boxes. They are then both examined to determine if they are to be kept or disregarded. If a box is not valid, it is eliminated. If it is valid, it is re-divided into two and so on and so forth until the boxes are of the required precision.

Our sensor works in the same way as a goniometre. In other words, sensorial data represents the observation angles of the environment's vertical landmarks. This means that they can not be linked to other elements on the map (such as horizontal landmarks). This is an advantage as it necessarily decreases the amount of matching combinations.

The localisation of a mobile robot using the theory of interval analysis has, of course, already been achieved, *e.g.* with telemetric sensors [15][13]. In a parametric sense, it is easy to see that our sensorial data are of the same nature as telemetric data

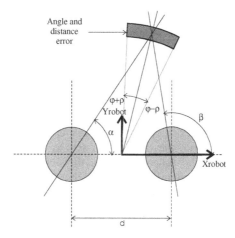

Fig. 3. Error modelisation Approach

used by M. Kieffer. Thus we have extrapolated the error model of Kieffer to our problem. This error model is characterised by both a distance and angular error.

At this level, we assume that our sensor provides the positions of the environment's vertical landmarks contaminated by an angular and a distance error. This forms an emission cone that resembles the one obtained by a telemetric sensor. The apex of this cone lies in the middle of the two images and on the axis that runs through their centre. As we know the angles α and β, we have the co-ordinates of the landmark, which enables us to calculate φ, the landmark's observation angle, and l, the distance measured (Figure 3).

If (x_r, y_r, θ_r) represents the robot's position, l the distance measured and φ the measured angle, then the computation of the co-ordinates of a point i on the map is calculated with the following formulas :

$$\begin{cases} x''_{si} = l_i \times \cos(\varphi_i) \\ y''_{si} = l_i \times \sin(\varphi_i) \end{cases} \qquad (2)$$

We then apply a rotation in the robot's reference frame that is equal to the orientation θ_r of the robot, followed by a switch from the robot's reference frame to the world's reference frame. Stating $[l_i]=[l_i - \varepsilon, l_i + \varepsilon]$ and $[\varphi_i]=[\varphi_i - \rho, \varphi_i + \rho]$ and using the inclusion functions $+, -, \times, \div, cos()$ and $sin()$ relative to the interval analysis, we obtain the following inclusion function:

$$\begin{aligned} [S_i] &= \begin{pmatrix} [x_{si}] \\ [y_{si}] \end{pmatrix} \\ &= \begin{pmatrix} \cos([\theta_r]) & -\sin([\theta_r]) \\ \sin([\theta_r]) & \cos([\theta_r]) \end{pmatrix} \times \begin{pmatrix} [l_i] \times \cos([\varphi_i]) \\ [l_i] \times \sin([\varphi_i]) \end{pmatrix} + \begin{pmatrix} [x_r] \\ [y_r] \end{pmatrix} \qquad (3) \\ &= f'([l_i],[\varphi_i],[x_r],[y_r],[\theta_r]) \end{aligned}$$

It is this inclusion function that will be used with the SIVIA algorithm.

First of all, once this box $[S_i]$ that corresponds to a sensorial data is found, we need to test if one of the map's elements is actually in this box. Given the fact that we are trying to estimate the position of the environment's vertical landmarks using a subpaving in order to obtain the imprecision, a landmark j of the environment is not represented by a point P_j, but by a subpaving made out of n boxes, that we note down as $[\![P]\!]_j = \{[T]_r / 1 \le r \le n\}$.

In order to obtain the Boolean inclusion function which will allow us to possess a global validity test, we apply this algorithm to the total of the sensorial data. The inclusion function used by SIVIA during this stage can be explained in the following way: For each localisation's box, we calculate if there is an intersection between the considered observation and the rectangular-set to be tested. As soon as the intersection is non-void, the function returns the undetermined value. During the initialisation of the map, the algorithm is limited, because there is no box representing the robot's localisation. Thus, we immediately have the subpaving that corresponds to the observation.

This situation only represents the case where there are no aberrant data. As a matter of fact, the algorithm successively tests all the boxes associated to the sensorial data and if only one is not valid, neither is the robot's position. Evidently, this situation presents several problems as it is quite common to have several aberrant data per acquisition.

Our solution to the problem is the same as the one adopted by M. Kieffer. It implements the algorithm whilst taking into account that there are no aberrant data. If no solution is found, the algorithm is repeated with one aberrant data, then two, etc. This solution gives a result no matter the ratio of "aberrant data /valid data".

In this case, the boxes are always divided until the minimal size that is defined by the error has been attained. We solely have the exterior approximation of the robot's position. Nevertheless it is the unique information that we are interested in for ulterior processing-computations that we will implement.

When using SIVIA, the first step is the search for a solution from a box received as an argument that has to contain the real position of the robot. One solution is to initialise this root box using the dimensions of the environment. The problem we are faced with however, is that on the one hand the calculation time is higher and on the other hand, in relatively symmetrical environments, the solution can be multiple and may not even contain the real position of the robot.

Bearing these facts in mind, we decided to use dead-reckoning information to refine the search for solutions. One method to use this information is based on the same principle as Kalman filtering, *i.e.* using successive phases of prediction/correction. This method works but needs specific algorithms that use subpaving and binary trees to compute the predicted state. Furthermore, modified SIVIA versions need to be used to take these particularities into account.

Having the most precise prediction phase as possible is very useful when the number of boxes is relatively high, as in the case of use of telemetric sensors. In our case, sensorial data represent the vertical landmarks of the environment and therefore the imprecision will be smaller and the number of boxes will be relatively low (as can be seen from the experimental results).

This is why we decided to only use dead-reckoning in order to initialise the initial box P_0 that is used to start the search for the robot's actual positions. From a rectangular-subpaving that results from a localisation process, we compute the minimal box that draws round the subpaving. This box is then increased with the maximum dead-reckoning error, which is a function of the distance covered.

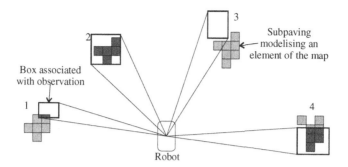

Fig. 4. Intersection test used in the localisation algorithm

This method is purely an initialisation phase and as we raise the dead-reckoning error, this implies that the possible results, which are incompatible with the actual position of the robot do not need to be tested during the localisation process.

3.2 Modelisation of the Environment

The representation of the data on the map is at the base of the SLAM paradigm. In our case, we need to focus on landmark's representation that is first of all compatible with the set interval analysis formalism and furthermore easy to use in an update phase. At this stage, the only solution that seems possible is a representation in subpaving.

The result of the localisation stage being a subpaving $[[L]]$, we can compute for each box $[L]_g$ (element of $[[L]]$) and for each sensorial data $[\varphi]_i$ and $[l]_i$, the box resulting in $f^1\left([L]_g,[\varphi]_i,[l]_i\right)$ thanks to the inclusion function that was already used in the localisation algorithm.

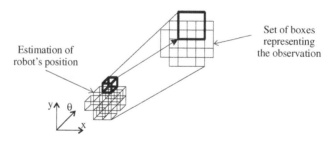

Fig. 5. Representation of all the rectangular-sets characterising a sensorial data

If we apply this inclusion function to the total of boxes rendered by the localisation stage, at the end of this process we obtain a set of boxes that correspond to each observation that can have a non-void mutual intersection and, therefore, do not constitute a subpaving (Figure 5).

This problematic has already been broached by [13]. As a matter of fact, he developed the *ImageSP* algorithm, which auto-decomposes into three phases, just to be able to calculate the image of a subpaving :

- **Hashing :** Calculates a regular subpaving $[\![A]\!]$ of which all the boxes have a size that is inferior to ε,
- **Evaluation :** Calculates the image of each of these boxes using the <u>considered</u> inclusion function f^I,
- **Regularisation :** Approximation of the union of these boxes $f^I([\![A]\!])$ using a new subpaving $[\![B]\!]$.

The first phase (*Hashing*) is unnecessary, given the fact that the subpaving $[\![L]\!]$ obtained during the localisation process is already made up solely of boxes that are smaller than the expected precision.

Thanks to the former inclusion function, we can directly compute the resulting box for each of these boxes and for all the sensorial data in the *evaluation* phase.

Finally, *the Regularisation* consist in using the new algorithm SIVIA to obtain the desired subpaving. The representation that we chose to use is a set of boxes of identical size, equal to the fixed minimal precision that characterises the two preceding sets. The advantage of this representation is that no bisection will be necessary when we need to process such a set. The boxes will be either accepted or rejected, as they are all of an inferior size to the expected precision. Using this method simplifies the representation of data in the map, but also the calculations that will be applied in the following phases. This may cause some problem if we want a very small precision but it is not applicable with the use of this sensor. Another method would be to use the exterior and interior approximation (Figure 6).

We now need to determine the inclusion function that will be used by the SIVIA algorithm during the addition of a new landmark in the map. As we want to obtain the set of boxes of an inferior size to the expected precision, this function should never render anything but two values: "true" or "undetermined". As a matter of fact, a "true" value rendered by this inclusion test would immediately stop the pending bisection of the box.

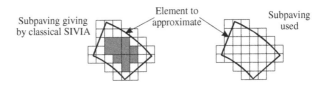

Fig. 6. Approximation of a set using the two former methods

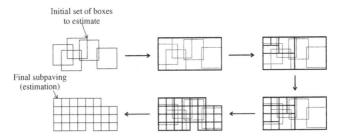

Fig. 7. Approximation of a set of rectangular-sets using SIVIA

This inclusion function plays a double role because it will be used to initialise the environmental map using the data issued from the first acquisition but also each time a new landmark is added to the map.

These two possibilities force us to differentiate between the two applications of this inclusion function. As a matter of fact, the robot's position is not a subpaving but a position during the initialisation phase, as it represents the origin of the map. However, when a new landmark is added the robot's position is defined by a set of boxes issued by the localisation phase. We will explain in detail our inclusion function in this second, more complicated case.

At this stage, we have to remind the reader that, of course, the direct observation image from a subpaving issued from the localisation phase provides a set of boxes, but not necessarily disjointed. This means that we need to estimate it, using a more practical and representative subpaving. Its only intersections' zones are the boxes' borders. In order to compute this set, we will again use the SIVIA algorithm: starting with an initial box, this will provide us the required subpaving. This will allow us to estimate each new landmark to be inserted in the map. Therefore, before running SIVIA, we need to compute an initial box. This can easily be done when calculating the minima and maxima from an observation for each box (Figure 7).

3.3 Decision Method for Matching

3.3.1 Determination of the Belief Put in Each Association

We now need to determine which information will have to be merged and which will have to be added to the map as new primitives. The decision method used here consists in determining a belief for each association, using the Dempster-Shafer theory [6][16]. This part of the process is crucial and decisive in the localisation paradigm and the simultaneous modelisation. In fact, it is this stage that will condition the maintenance of the environmental map's coherence. A wrong choice between a new insertion or fusion will generally be at the root of an excess of primitives in the map, which will lead to cumulative errors and, hence, a divergence in the algorithm.

At the start of this phase, we have three imprecision's data at our disposal that will be uses:

- An environmental map made up of subpaving each representing the imprecision associated to the modelled landmark.
- A set of sensorial data characterised by information of the distance/angle type in the form of intervals, providing the imprecision in the measure,
- A subpaving resulting from the localisation stage, representing the imprecision associated to the robot's position.

We therefore have to resolve two principal problems:
- Define and use the set resulting from the association of the localisation and the measure imprecision;
- Find a comparison criteria that can be implemented to determine the belief attributed to the fusion of this set with a map's subpaving.

These two problems are tightly linked and in order to know if an observation can indeed be associated to a mapped primitive, we need to find a comparison criteria between the two: the intersection of the two subpavings. In fact, the more the set associated to an observation contains the subpaving that represents a point on the map, the more certain we are that it represents the same information, which implies that they have to be merged.

Given the fact that the set of boxes of an observation can overlap, several of them can have a non-void intersection with one of box representing a point on the map. This is why we cannot directly use the intersection notion between these different boxes to calculate the volume. In fact, if we were to consider the three sets A, B and C so that $A \cap B \cap C \neq \emptyset$, we would obtain the following inequation:

$$\text{Volume}(A \cap C) + \text{Volume}(B \cap C) > \text{Volume}(A \cap B \cap C) \qquad (4)$$

This signifies that the volume that corresponds with the intersection of the sets A and B with C is counted twice in the left part of the inequation. The chosen solution is then the same as when adding a new observation to the map. In other words, we calculate the image of the subpaving issued from a localisation and then SIVIA is applied to obtain a subpaving associated to the observation that we note as $[[S]]_i =$ $\{[K]_q / 1 \leq q \leq m\}$ with $1 \leq i \leq s$ and m representing the amount of boxes constituting the subpaving.

Our comparison criterion is therefore based on the value:

$$\tau = \left(\text{Volume} \left([[P]]_j \right) - \text{Volume} \left([[P]]_j \cap [[S]]_i \right) \right) \times 100, \qquad (5)$$

that represents the percentage of $[[P]]_j$ included in $[[S]]_i$.

The formalism used to determine the certainty associated to a fusion is based on the search of the maximum of belief compared to the application of the Dempster-Shafer rules. We therefore have to determine our discernment-frame constituted out of two elements: $\Theta = \{\text{YES, NO}\}$

- "YES" the observation i needs to be merged with the element j on the map
- "NO" the observation i should not be merged with the element j on the map

If the subpaving issued from an observation contains more than 50 % of the boxes that define a landmark, we consider that the belief must be the highest. Thus, we use the Basic Probability Assignment (B.P.A.) as represented in figure 8.

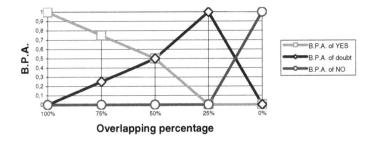

Fig. 8. Matching functions for the fusion stage

All we now need to do is to compute the intersection volume that exists between each subpaving $[[S]]_i$ issued from an observation and each subpaving $[[P]]_j$ that represents a landmark on the map.

For an observation S_i, we now have p triplets:

$$\begin{array}{lll} m_{i,1}(P_1) & m_{i,1}(\overline{P_1}) & m_{i,1}(\Theta_1) \\ m_{i,2}(P_2) & m_{i,2}(\overline{P_2}) & m_{i,2}(\Theta_2) \\ \dots & \dots & \dots \\ m_{i,p}(P_p) & m_{i,p}(\overline{P_p}) & m_{i,p}(\Theta_p) \end{array}$$

We can now compute these p triplets for the s observations, which will give us $s \times p$ triplets. The problematic introduced at this level resides in the fusion of all the information, in order to be able to choose. We resolved this problem by using the generalisation of the combination operator of Dempster-Shafer introduced by D. Gruyer and V. Cherfaoui [10].

3.3.2 Decisional Algorithm

The decisional algorithm that we use is based on the maximum of the probability obtained in the Dempster-Shafer sense. The precedent phase allowed us to calculate for each observation, p triplets that correspond to the match with each element on the map. We can now apply the generalised Dempster-Shafer operator in order to obtain a matrix of belief with the dimensions $s \times (p+2)$. The hypothesis "*" signifies that the observation S_i does not correspond with any element on the map. This means we work in a extended open world.

The result of our matrix of belief provides a belief onto the singletons hypothesis, *i.e.* a rule of decision based on the maximum pignistic probability will not add anything here because this last one use a group of elements. Furthermore, the values of this matrix are directly credibilist measures. This is why we have based our decisional criterion on the maximum credibility of this matrix.

The algorithm used is based on the search for the maximum value in the matrix previously built. The value that is found this way allows us to determine if the observed point is in relation with an existing point or if a new point has been created. In case of doubt (maximum credibility on "Θ"), we choose to create a new point defined by a subpaving.

Once this match has been carried out, all the elements of the line that contain the maximal value are put on 0, as well as those of the colon but only if this last one is different from "*"and from "Θ". In fact, the initialisation of all the elements of the line to 0 signifies that an observed element cannot be in relation to one single element on the map. The same applies for the colon that corresponds to the fact that several observations cannot be matched to the same point on the map. On the other hand, several observations can be new points ("*") just like the ignorance can be maximal in several ("Θ") observations. The algorithm is reiterated as long as there are positive values.

Finally, this algorithm gives us two sets. The first is made up of observations that need to be merged with an element on the map. The second is made up of new landmarks that need to be added. The processing and management of these two sets will be presented in the next part.

3.4 Incremental Update of the Environment's Map

The former decisional integration/fusion stage, has provided us with two sets of points: the first contains those that need to be merged and the second those that need to be added to the map. The integration of a new element on the environment's map has already been given previously.

The last stage that needs to be processed is the fusion between an element from the map and an observation. Here, the data are defined by sets and as we find ourselves in a context of bounded error, the actual position of the landmark has to belong to the two sets. The result of the fusion of an observation with an element of a map is therefore the intersection of the two sets.

At this level we need to resolve a problem. In fact, each set is defined by several boxes. The one that represents the observation even contains boxes that can overlap. The calculation of the intersection is brought back to processing the problem of multiple intersections of disjointed boxes. It is far from a trivial problem.

In order to overcome this difficulty, we part from the following fact: as the solution belongs to both sets, one of the two can first be considered. Then, we can check if each box from the first set, is an element of the second set. If this is the case the box is kept, otherwise it is eliminated. The set of boxes most adapted to be the first set is then the one that represents the landmark on the map, as it is uniquely made of separate boxes, *i.e.* a subpaving (Figure 9).

We can observe at this point that the result of our fusion method can only contain a reduction of subpaving representing the imprecision of a landmark on the map. No matter the set associated with the observation, after fusion there can only be an addition of information in the sense that the subpaving of the landmark cannot increase.

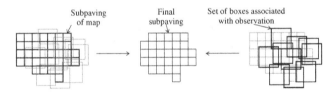

Fig. 9. Example of fusion between observation and element on the map

In order to validate our approach, we present the experimental results in the next part. These results were obtained in two distinctive environments.

3.5 Experimental Results

We have tested our SLAM method in two types of structured environments.

The first series of 8 acquisitions has enabled us to validate our paradigm of localisation and simultaneous modelling in a small environment.

The second series of measures contains 45 acquisitions realised during a trajectory consisting of a return trip. The distance covered is approximately fourteen meters. Here, we use dead-reckoning to reduce the size of the box that looks for the possible positions of the robot. We remind the reader that the dead-reckoning error is maximised in order to serve uniquely in the initialisation of the SIVIA algorithm.

Certain elements of the map can be eliminated as we go along updating the localising. In fact, the filtering that we have developed and that we use here, allows us to keep the elements in the map that have already been observed several times.

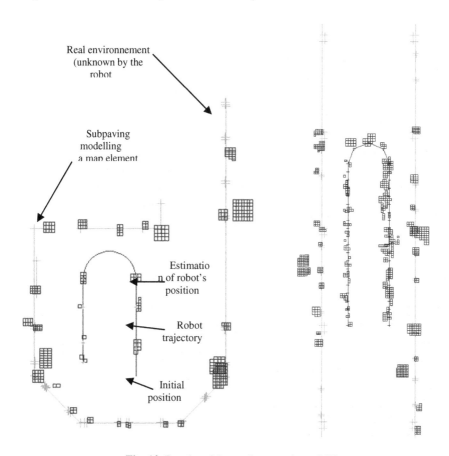

Fig. 10. Results of the environment's modelling

First, from a general viewpoint, the simultaneous process of localisation and modelling provides coherent results in terms of precision and in terms of robustness. Furthermore, we can see that the SLAM process does not diverge. In fact, the 45 acquisitions result in a coherent construction of the environmental map with no preliminary knowledge.

From a localisation viewpoint, we can affirm that the absence of the preliminary knowledge had not effected the estimation phase of the robot's configuration using interval analysis. The coherence of the localisation phase is also proved by the variation during the movement of the robot. We see that the subpaving decreases on the way back (in other words after the U-turn) than on the way there. From a modelling viewpoint, and still linked to the observations of a general order, we can affirm that the map generated is coherent in comparison with the actual terrain. The amount of cartographic integrated primitives is coherent, proving the validity of the fusion and integration process.

The evolution of the subpavings during the incremental modelling process is robust and coherent. Two points can justify this statement: First of all, the contribution of sensorial data is accompanied by a reduction of the size of the error domain and by a convergence of subpaving to the actual position of landmarks. Secondly, the interaction between the localisation error and the modelling error is taken into account because the higher the localization precision, the more the subpavings on the cartographic primitives are significantly reduced. This decisive factor allows the process of simultaneous localisation and modelling not to diverge after a certain amount of acquisitions. This test on large environments is important as it is put forth by several works, such as those of Dieter Fox [9].

Rather than an alternative, the interval analysis approach is proposed as a solution that allows us to integrate intrinsically the imprecision notion. The fact that we can manage the imprecision implies the possibility to take the interactions into account, which is not possible with other formalisms. It is this rigorous management of these interactions that leads to a successful outcome of the map process generation on long distances.

4 Conclusion

In this work, we have developed a method of localisation and simultaneous modelling (SLAM) of the environment based on the use of the interval analysis. This method is different from classical algorithms found in literature and that are generally probabilistic. The novelty of the proposed formalism resides in the fact that the obtained imprecision domains linked to the state's estimation are equiprobable and guaranteed.

We have given preference to the use of the Dempster-Shafer rules, that allow us to manage a belief in different cases that can appear in a map-generation process from each observation (fusion, insertion or rejection).

The strategy to integrate primitives carried over is the reduction of the subpaving matched to the examined and mapped primitive. This technique processes rapidly but first needs all the elements to be inserted as subpavings reduced to the minimum. In other words, the size of each box has to be inferior to the expected precision.

We have seen that the method developed provides excellent results. First of all the paradigm, validated on a trajectory in a long corridor, gives a high precision on the localisation and the estimation of the landmarks' position. Secondly, no localisation drift has been observed.

Here we have a system that can simultaneously localise the robot from a non-reliable map and at the same time incrementally model the robot's evolution in the environment in a relatively precise way. These two stages being intimately linked, the quality of the one depends on the precision of the other. The use of the interval analysis has allowed us to propagate the imprecision introduced during each stage of our method on the next phases.

Bibliographie

1. J. Boreinstein, Y. Koren, "Histogrammic in-motion mapping for mobile robot obstacle avoidance", IEEE Trans. On rob. and auto., Vol. 7, N°4, pp. 1688-1693, August 1991

2. C. Cauchois, E. Brassart, C. Pegard, A. Clerentin – "Technique for Calibrating an Omnidirectional Sensor" – Proc. of the IEEE International Conference on Intelligent Robots and Systems (IROS'99), october 1999, p. 166-171.

3. H. Choset and J. Burdick, "Sensor Based Planning, Part I: The Generalized Voronoi Graph," Proc. of IEEE Int. Conf. on Rob. and Auto. (ICRA '95), Vol.2, pp. 1649-1655, May 1995

4. J. Crowley, "World modelling and position estimation for a mobile robot using ultrasonic ranging", Proc. Of IEEE Conference on Robotics and Automation, Scottsdale, May 1989, p. 674-680.

5. L. Delahoche, C. Pégard, B. Marhic, P. Vasseur - "A navigation system based on an omnidirectional vision sensor" - Proceedings on IEEE/RSJ International Conference on Intelligent Robots and Systems (IROS'97), Grenoble, France, Septembre 1997.

6. A. P. Dempster – "Upper and lower probabilities induced by a multi-valued mapping" – Annals of Mathematical Statistics, vol. 38, 1967.

7. C. Drocourt, L. Delahoche, C. Pégard, C. Cauchois. "Localisation method based on omnidirectional stereoscopic vision and dead-reckoning" - Proc. of the IEEE International Conference on Intelligent Robots and Systems (IROS'99), Korea, pages 960-965, October 1999.

8. A. Elfes, "Sonar-based real world mapping and navigation", IEEE Journal of robotics and automation, Vol. RA-3, N°3, pp. 249-265, June 1987.

9. D. Fox, W. Burgard, S. Thrun, "Probabilistic Methods for Mobile Robot Mapping", Proc. of the IJCAI-99 Workshop on Adaptive Spatial Representations of Dynamic Environments, 1999.

10. D. Gruyer, V. Berge-Cherfaoui – "Matching and decision for Vehicle tracking in road situation" – IEEE/RSJ International Conference on Intelligent Robots and Systems, IROS'99, Kyongju, Corée, 17-21 octobre 1999.

11. J. Guivant, E. Nebot, S. Baiker, "Autonomous navigation and map building using laser range sensors in outdoor applications", Journal of Robotic Systems, Vol 17, n° 10, pp 565-283, October 2000.

12. L. Jaulin, and E. Walter – "Global numerical approach to nonlinear discrete-time control" – IEEE Trans. on Autom. Control, 42, 872-875 (1997).

13. M. Kieffer, L. Jaulin, E. Walter and D. Meizel – "Localisation et suivi robustes d'un robot mobile grâce a l'analyse par intervalles" – Traitement du signal, volume 17, n° 3, 207-219.

14. B. Kuipers, Y.T. Byun, « A robot exploration and mapping strategy based on a semantic hierarchy of spatial representations », Robotics and Autonomous Systems, 8 1991.

15. O. Leveque – "Méthodes ensemblistes pour la localisation de véhicules" – Thèse de doctorat, Université de Technologie de Compiègne, 1998.

16. G. Shafer - "A mathematical theory of evidence" - Princeton : university press, 1976

17. P. Smets et R. Kennes – "The transferable belief model" – Artificial Intelligence, vol. 66 n°2 pages 191-234, 1994.

18. Y. Yagi, S. Kawato - "Panorama Scene Analysis with Conic Projection" - IEEE International Workshop on Intelligent Robots and Systems, 1990, p 181-187.

Nonlinear Predictive Control Using Constraints Satisfaction

Fabien Lydoire and Philippe Poignet

Laboratoire d'Informatique,
de Robotique et de Microélectronique de Montpellier,
UMR CNRS UM2 5506, 161 rue Ada,
34392 Montpellier Cedex 5, France
{lydoire, poignet}@lirmm.fr

Keywords: interval analysis, state estimation, nonlinear model predictive control.

1 Introduction

During the last few years, control schemes using interval analysis have been investigated. Several approaches have been proposed in order to get robust control in presence of model uncertainties [7, 10] or for state estimation [6].

In this paper, we investigate the design of a nonlinear model predictive controller [1], using set computation. The motivation for using NMPC control is its ability to handle nonlinear multi-variable systems that are constrained in the state and/or in the control variables. The NMPC problem is usually formulated as a nonlinear constrained optimisation one, and is solved using classic non linear optimisation techniques. However, most of the NMPC constraints are easily expressed using intervals. Therefore, we will use interval analysis techniques [8] in order to compute an NMPC constraints satisfying solution. Classic interval branch and bound algorithms have been investigated for predictive control in [3]. They conclude that the pessimism introduced by interval computation in the estimation of the states leads to high computational cost and may only be used on control of low dynamic systems. Therefore, we propose a new approach based on a spatial discretisation of the input and state domains to improve interval model predictive control and to be applied on high dynamic systems. The proposed strategy will be numerically simulated on an inverted pendulum model.

The paper is organised as follows : section 2 presents the classical nonlinear model predictive control technique, section 3 introduces interval analysis, set inversion and the proposed algorithm for its application to the NMPC problem. Finally section 4 exhibits numerical simulation results.

2 Nonlinear Model Predictive Control

The NMPC problem [1] is usually formulated as a constrained optimization problem

C. Jermann et al. (Eds.): COCOS 2003, LNCS 3478, pp. 142–153, 2005.

$$\min_{\boldsymbol{u}_k^{N_p}} J(x_k, \boldsymbol{u}_k^{N_p}) \tag{1}$$

subject to

$$x_{i+1|k} = f(x_{i|k}, u_{i|k}) \quad x_{0|k} = x_k \tag{2}$$
$$u_{i|k} \in \mathbb{U}, \quad i \in [0, N_p - 1] \tag{3}$$
$$x_{i|k} \in \mathbb{X}, \quad i \in [0, N_p] \tag{4}$$

where

$$\mathbb{U} := \{u_k \in \mathbb{R}^m | u_{\min} \leq u_k \leq u_{\max}\}$$
$$\mathbb{X} := \{x_k \in \mathbb{R}^m | x_{\min} \leq x_k \leq x_{\max}\} \tag{5}$$

Internal controller variables predicted from time instance k are denoted by a double index separated by a vertical line where the second argument denotes the time instance from which the prediction is computed. $x_k = x_{0|k}$ is the initial state of the system to be controlled at time instance k and $\boldsymbol{u}_k^{N_p} = [u_{0|k}, u_{1|k}, \ldots, u_{N_p-1|k}]$ an input vector.

Predictive control (fig. 1) consists on computing the vector $\boldsymbol{u}_k^{N_p}$ of consecutive inputs $u_{i|k}$ over the prediction horizon N_p and applying only the solution input $u_{0|k}$. These computations are updated at each sampling time.

The dynamic model of the system is written as a nonlinear equality constraint on the state (eq. 2). Bounding constraints over the inputs $u_{i|k}$ and the state variables $x_{i|k}$ over the prediction horizon N_p are defined through the sets $\mathbb{U}$ and $\mathbb{X}$ (eq. 5).

The objective function J is usually defined as

$$J(x_k, \boldsymbol{u}_k^{N_p}) = \phi(x_{N_p|k}) + \sum_{i=0}^{N_p-1} L(x_{i|k}, u_{i|k}) \tag{6}$$

where ϕ is a constraint over the state at the end of the prediction horizon, called state terminal constraint, and L a quadratic function of the state and inputs.

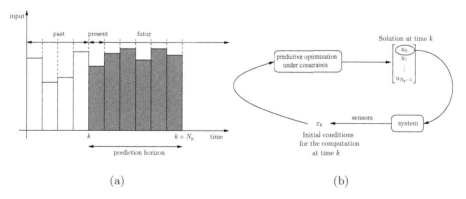

(a)   (b)

Fig. 1. Principles of the predictive constrained optimal control approach

The solution $u_k^{N_p}$ of the NMPC problem has two properties. Firstly, it satisfies the constraints over the inputs ($u_k \in \mathbb{U}$) and the states ($x_k \in \mathbb{X}$), including the state terminal constraint. Secondly it is optimal with respect to the criteria J. In this article, we will consider the computation of a solution satisfying the constraints, without considering the optimisation.

Except for the dynamic model of the system (eq. 2) which is nonlinear, NMPC constraints (eqs. 3,4) are inequality constraints and can directly be written as intervals. Therefore, it would be interesting to use interval techniques in order to compute a solution satisfying the NMPC constraints. The following section introduces interval analysis concepts used to compute such a solution.

3   Constraints Satisfaction

3.1   Interval Analysis and Set Inversion

Initially dedicated to finite precision arithmetic for computer [11] and after used in a context of guaranteed global optimization [4], the interval analysis is based on the idea of enclosing real numbers in intervals and real vectors in boxes.

Let f be a function from $\mathbb{R}^n$ to $\mathbb{R}^m$ and let $\mathbb{Y}$ be a subset of $\mathbb{R}^m$. Set inversion is the characterization of

$$\mathbb{X} = \{x \in \mathbb{R}^n \mid f(x) \in \mathbb{Y}\} = f^{-1}(\mathbb{Y}) \tag{7}$$

Set inversion algorithms [8] are based on consecutive bisections of an initial domain $[x]$ for $\mathbb{X}$. They can perform inner ($\underline{\mathbb{X}}$) and outer ($\overline{\mathbb{X}}$) approximation of $\mathbb{X}$ ($\underline{\mathbb{X}} \subset \mathbb{X} \subset \overline{\mathbb{X}}$). The image $f([x])$ of $[x]$ is computed and compared to $\mathbb{Y}$. Four cases may be encountered:

1. $f([x]) \cap \mathbb{Y} = \emptyset$, then $[x]$ is rejected as a subset of $\mathbb{X}$ (fig. 2(b)).
2. $f([x]) \subset \mathbb{Y}$, then $[x]$ is a subset of $\mathbb{X}$ and therefore $[x]$ is stored into $\underline{\mathbb{X}}$ and $\overline{\mathbb{X}}$.
3. $f([x]) \not\subset \mathbb{Y}$ and $f([x]) \cap \mathbb{Y} \neq \emptyset$, then $[x]$ may contain a part of the solution set. If its width is greater than a precision threshold ϵ, then $[x]$ is bisected and the test is recursively applied (fig. 2(b)).
4. If the test gives the same results as in case 3, and if the width of $[x]$ is lower than ϵ, then $[x]$ is stored into $\overline{\mathbb{X}}$.

Figure 2(c) illustrates the inner approximation of $f^{-1}(\mathbb{Y})$ finally computed by the set inversion algorithm.

Considering the initial domain $[x_{\min}, x_{\max}]$, the algorithm brackets the solution set $\mathbb{X}' = [x_{\min}, x_{\max}] \cap f^{-1}(\mathbb{Y})$ by two subpavings $\underline{\mathbb{X}}$ and $\overline{\mathbb{X}}$.

$$\mathbb{X}' = \{x \in [x_{\min}, x_{\max}] \mid f(x) \in \mathbb{Y}\} \subseteq f^{-1}(\mathbb{Y}) \tag{8}$$

3.2   Application to the NMPC Problem

The purpose is to apply the set inversion algorithm to compute a solution satisfying the NMPC constraints.

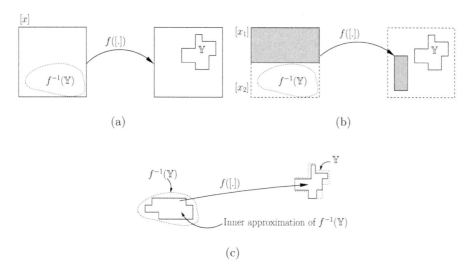

(a)   (b)

(c)

Fig. 2. Set inversion algorithm steps

Considering the set inversion formulation, $\mathbb{Y}$ domains are defined by the limits over the state variables, and the initial domain which will be bisected during the algorithm is defined by the limits over the inputs (eq. 5).

The dynamic model function f is applied over the horizon starting from the current state x_k. The computation of a new state domain $[x_{i+1}]$ from previous state domain $[x_i]$ and input domain $[u_{i_{\min}}, u_{i_{\max}}]$ is followed by the set inversion algorithm (fig. 3).

This procedure bisects the initial domain $[u_{i_{\min}}, u_{i_{\max}}]$ and provides a domain $[u_i]$ such that

$$f([x_i], [u_i]) = [x_{i+1}] \text{ and } [x_{i+1}] \subseteq [x_{i+1_{\min}}, x_{i+1_{\max}}] \tag{9}$$

where $[x_{i+1_{\min}}, x_{i+1_{\max}}]$ is the feasible domain for the state x_{i+1} (eq. 4).
The bisection procedure reducing the width of an interval, $[u_i]$ is such that

$$[u_i] \subseteq [u_{i_{\min}}, u_{i_{\max}}] \tag{10}$$

and therefore any punctual value in the interval $[u_i]$ is a solution satisfying the NMPC constraints.

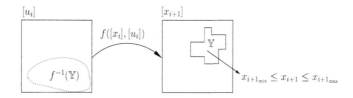

Fig. 3. Set inversion algorithm applied on NMPC

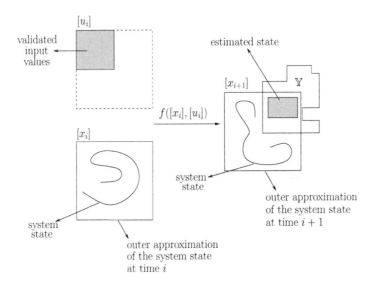

Fig. 4. Validation of incorrect input due to outer approximation of the state

The computation of an input satisfying the NMPC constraints implies the state estimation of the system with interval values (eq. 9). State estimation involves the computation of the dynamic model of the system followed by an integration and therefore introduces pessimism in the estimation of the states domains. State estimation on intervals are based on interval Taylor series [5, 12] and lead to guaranteed but outer approximation of the system state. Therefore the intersection of the computed state with the state constraints during the set inversion algorithm may be composed of outer state values. Consequently, the input is validated by the set inversion algorithm whereas it does not satisfy the NMPC constraints (fig. 4).

In the following, we will propose a solution to get an inner approximation of the state and thus use the set inversion algorithm to compute a NMPC constraints satisfying solution.

3.3   NMPC Constraints Satisfaction

Classical state estimation over intervals leads to outer approximation. However the preceding section exhibited the need for an inner approximation of the system state. Therefore, we will compute state estimation over the horizon using punctual values distributed in the considered domains.

In the following, we will omit the index $_{|k}$ assuming that prediction is made at time instance k.

On each iteration, the set of inputs $u_i^1, \ldots, u_i^n$ which define a spatial distribution of the input domain $[u_{i_{\min}}, u_{i_{\max}}]$, is applied on each punctual state values $x_i^1, x_i^2, \ldots, x_i^m$ defining a spatial discretisation of $[x_i]$. This gives a new set of punctual values defining a spatial discretisation of $[x_{i+1}]$ (fig. 5).

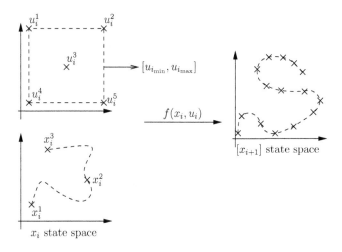

Fig. 5. Spatial discretisation

Assuming that f is continuous, the spatial discretisation of $[x_{i+1}]$ computed by the algorithm provides an inner approximation of $f([x_i], [u_{i_{\min}}, u_{i_{\max}}])$. Indeed, for any punctual value x_i^p in $[x_i]$, $p \in [1, m]$, and any inputs u_i^l and u_i^{l+1}, $l \in [1, n-1]$ continuity of f leads to

$$[\min(f(x_i^p, u_i^l), f(x_i^p, u_i^{l+1})), \max(f(x_i^p, u_i^l), f(x_i^p, u_i^{l+1}))] \subseteq f(x_i^p, [u_i^l, u_i^{l+1}]) \tag{11}$$

therefore the set of input variables $\mathbb{S}'$ considering the inner approximation of the state

$$\begin{aligned}
\mathbb{S}' = \{ u_i \in [u_i^l, u_i^{l+1}] \mid \\
[\min(f(x_i, u_i^l), f(x_i, u_i^{l+1})), \max(f(x_i, u_i^l), f(x_i, u_i^{l+1}))] \\
\subseteq [x_{i+1_{\min}}, x_{i+1_{\max}}] \}
\end{aligned} \tag{12}$$

is an inner approximation of the set of input variables $\mathbb{S}$ in case of perfect state estimator over intervals.

$$\mathbb{S} = \{ u_i \in [u_i^l, u_i^{l+1}] \mid f(x_i, [u_i^l, u_i^{l+1}]) \subseteq [x_{i+1_{\min}}, x_{i+1_{\max}}] \} \tag{13}$$

The inner approximation of the state of the system allows the use of the set inversion algorithm to compute a solution satisfying the NMPC constraints. The efficiency of the solution depends on the sampled values u_i^l of the initial input interval $[u_{i_{\min}}, u_{i_{\max}}]$, and on the accuracy threshold ϵ defining the minimum width for an interval allowed to be bisected during the set inversion procedure.

One of the drawback of the inner approximation of the state is that state values outside the inner approximation are not considered and therefore could violate the constraints (fig. 6). This leads to the validation of an incorrect input domain. However, the punctual values defining the spatial discretisation of the state are guaranteed to belong to the constrained space. These values have been

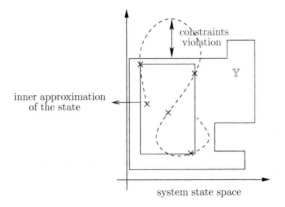

Fig. 6. Constraints violation due to the inner approximation of the state

computed from punctual input values defining the spatial discretisation of the input domain. Therefore, theses punctual input values are guaranteed to lead to the constrained state space. However, picking any punctual value in the computed input interval may lead to constraint violation. This constraint violation has not been characterized yet and will be the object of future work.

4   Simulation Results

The control scheme presented in this paper is applied on the stabilisation of an inverted pendulum. The pendulum is free to rotate around an horizontal axis and is actuated by a linear motor whose acceleration is the input of the system. Friction has been neglected and the hypothesis is made that the pendulum is a rigid body.

Let's consider the inverted pendulum (fig. 7) which is a classical benchmark for nonlinear control techniques [2, 9]. Its dynamic equation (eq. 2) where $x = [q, \dot{q}]^T$ is based on the following equation

$$\ddot{q}_{t+1} = K_{\sin} \sin(q_t) - K_{\cos} u_t \cos(q_t) \tag{14}$$

Friction has been neglected and it has been assumed that the pendulum is a rigid body.

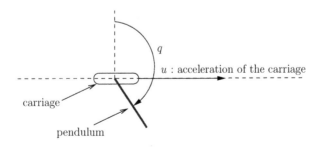

Fig. 7. The inverted pendulum

The acceleration $\ddot{q}_{t+1}$ is integrated twice using:

- first order Taylor series in the predictive controller,

$$\dot{q}_{t+1} = \dot{q}_t + \delta_t \ddot{q}_{t+1} \tag{15}$$

$$q_{t+1} = q_t + \delta_t \dot{q}_{t+1} \tag{16}$$

where δ_t is the time sampling period
- Runge-Kutta formula in the simulator.

In the simulations, a single $[u]$ value is bisected over the horizon. Parameters $K_{\sin}$ and $K_{\cos}$ have been computed from a real pendulum available at the laboratory. The parameter nb_{samples} define the number of punctual values used in the spatial discretisation of $[u]$. N_p is the prediction horizon, the initial state is $[q_{\text{ini}}, \dot{q}_{\text{ini}}]^T$, the precision threshold used for bisection in the set inversion algorithm is ϵ. The feasible values are those defined by NMPC inequalities (eqs. 3,4). The common parameter values are regrouped in the following table

$K_{\sin}$	$K_{\cos}$	$\dot{q}_{\text{ini}}$ (rad.s^{-1})	δ_t (s)
109	11.11	0	0.001

$[q_{\text{feasible}}]$ (rad)	$[\dot{q}_{\text{feasible}}]$ (rad.s^{-1})	$[u_{\text{feasible}}]$ (m.s^{-2})
$[-\pi - \frac{3\pi}{2}; -\pi + \frac{3\pi}{2}]$	$[-150;150]$	$[-800;800]$

The punctual value u applied on the system is the closest to zero in the solution interval.

The simulations have been computed using MATLAB with a 2Ghz PENTIUM IV.

In simulations 4.1 to 4.2, the computation of the domain $[u]$ is stopped as two valid punctual values defining the spatial discretisation of $[u]$ have been determined. In simulations 4.4, the computation of $[u]$ is achieved completely.

4.1   Initial Position Downwards

This simulation has been executed with the initial position downwards.

N_p	q_{ini} (rad)	ϵ (m.s^{-2})	$[q_{N_p}]$ (rad)	nb_{samples}
40	$-\pi$	1.0	$[-0.1;0.1]$	5

Figure 8 displays the results of this simulation. The pendulum starts from initial position $-\pi$ and is stabilised by the control law in its terminal position $q = [-0.1; 0.1]$ rad.

4.2   Initial Position Close to 0 Rad

This simulation has been executed with the initial position close to the terminal position. It exhibits the computation time variation due to the reduction of the prediction horizon.

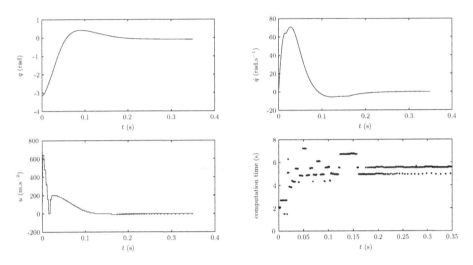

Fig. 8. Joint position, velocity, input and computation time of the input

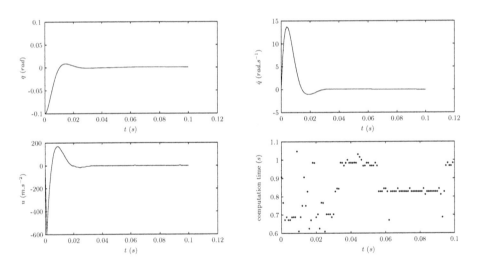

Fig. 9. Joint position, velocity, input and computation time of the input

N_p	q_{ini} (rad)	ϵ (m.s^{-2})	$[q_{N_p}]$ (rad)	nb$_{\text{samples}}$
5	−0.1	1.0	[-0.001;0.001]	5

Figure 9 displays the results of this simulation. As in the previous simulation, the pendulum is stabilised in its terminal position. However, the computation time is reduced by a factor ~ 6.

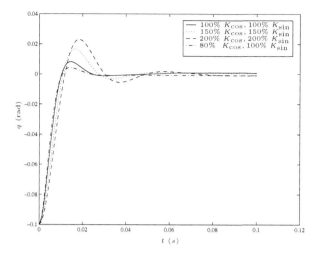

Fig. 10. Simulations with model parameters errors

4.3   Robustness with Respect to Model Error

The following simulations have been executed with the parameters used in simulation 4.2. Model error have been introduced through errors on the parameters K_{sin} and K_{cos}.

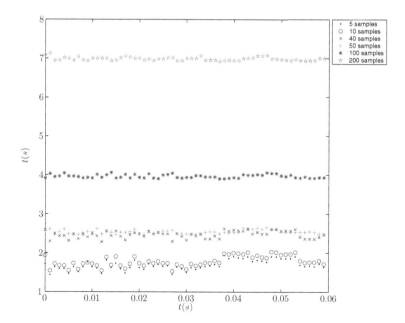

Fig. 11. Computation time with different $nb_{samples}$ values

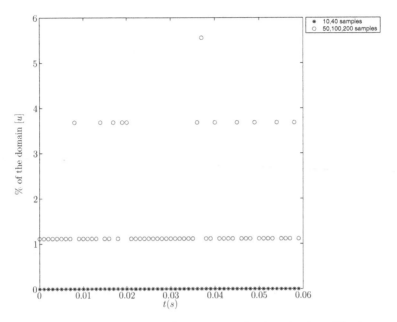

Fig. 12. Domain size percentage with respect to the domain size with $nb_{samples} = 5$

Figure 10 exhibits the robustness of the method by displaying the joint positions. In each presented case, the control method leads the pendulum to the final constrained position. In the case of a value inferior or equal of 70% of the exact model value for K_{cos}, the algorithm is unable to find a solution.

4.4   Spatial Discretisation Variation

The simulations presented in this section exhibit the influence of the parameter $nb_{samples}$ on the calculation of the domain $[u]$ and on computation time. The more samples there is, the longer is the computation time (fig. 11). However, the computed domain for $[u]$ is not increased a lot (fig. 12). This is due to the algorithm used. Whatever the number of samples, the domain will be bisected until the bisected domains will be too small ($< \epsilon$) to be bisected. Increasing the number of samples avoid bisections but introduces much more small domains to deal with.

5   Conclusion

This paper introduces a nonlinear control approach associated with interval analysis. The guaranteed state estimation techniques have been demonstrated to be inappropriate. Therefore, an inner bounding state estimation method for continuous systems has been presented. The complete simulation results show the efficiency and the robustness of the proposed method.

Future work will concern the following two points. Firstly, the computational efficiency improvement by taking into account contraction procedure based on constraints propagation. Secondly, the characterisation of the inner approximation of the state in order to compute input boxes satisfying the constraints completely.

References

1. Frank Allgöwer, Thomas A. Badgwell, S. Joe Qin, James B. Rawlings, and Steven J. Wright. Nonlinear predictive control and moving horizon estimation - an introductory overview. In P.M. Frank, editor, *Advances in Control: Highlights of ECC '99*, chapter 12, pages 391–449. Springer-Verlag, 1999.
2. Karl Johan Åström and Katsuhisa Furuta. Swinging up a pendulum by energy control. *Automatica*, 36:287–295, 2000.
3. J.M. Bravo, C.G. Varet, and E.F. Camacho. Interval model predictive control. In *IFAC Algorithm and Architectures for Real-Time Control*, 2000.
4. Eldon. R. Hansen. *Global Optimization Using Interval Analysis*. Marcel Dekker, New York, NY, 1992.
5. Kenneth R. Jackson and Nedialko S. Nedialkov. Some recent advances in validated methods for ivps for odes. *Applied Numerical Mathematics*, 42:269–284, 2002.
6. Luc Jaulin. Nonlinear bounded-error state estimation of continuous-time systems. *Automatica*, 38:1079–1082, 2002.
7. Luc Jaulin, Isabelle Braems, Michel Kieffer, and Eric Walter. Interval methods for nonlinear identification and robust control. In *Proceedings of the IEEE Conference on Decision and Control (CDC)*, Las Vegas, Nevada, 2002.
8. Luc Jaulin, Michel Kieffer, Olivier Didrit, and Eric Walter. *Applied Interval Analysis*. Springer, 2001.
9. Lalo Magni, Riccardo Scattolini, and Karl Johan Åström. Global stabilization of the inverted pendulum using model predictive control. In *Proceedings of the 15th IFAC World Congress*, Barcelona, 2002.
10. Stefano Malan, Mario Milanese, and Michele Taragna. Robust analysis and design of control systems using interval arithmetic. *Automatica*, 33(7):1363–1372, 1997.
11. Ramon E. Moore. Methods and applications of interval analysis. *Philadelphia, SIAM*, 1979.
12. Tarek Raïssi, Nacim Ramdani, and Yves Candau. Garanteed stated estimation for nonlinear continuous systems with taylor models. In *Proceedings of the IFAC Symposium on System Identification (SYSID)*, 2002.

Gas Turbine Model-Based Robust Fault Detection Using a Forward – Backward Test

Alexandru Stancu, Vicenç Puig, and Joseba Quevedo

Automatic Control Department - Campus de Terrassa,
Universidad Politécnica de Cataluña (UPC),
Rambla Sant Nebridi, 10. 08222 Terrassa (Spain)
{Alexandru.Stancu, Vicenc.Puig, Joseba.Quevedo}@upc.es

Abstract. The problem of robust model based fault detection of dynamic systems using interval observers has been mainly addressed checking if the measured output is inside the interval of possible estimated outputs obtained considering uncertainty on model parameters. This task can be computationally expensive because the interval observers can be affected by the wrapping effect. In this paper, a mixed approach consisting in determining a computationally cheaper inner approximation of the estimated output interval, based only on simulating vertices of parameter uncertainty region (forward test), is combined with a backward consistency check when the real measured output falls outside this inner solution (backward check). The backward check is implemented using interval constraint satisfaction algorithms which can perform efficiently in deciding if the measured output is consistent with the interval model. The classical alternative to this backward check will force to solve a global optimisation problem, or equivalently, a global consistency problem. Finally, this approach will be tested on a gas turbine nozzle servosystem.

1 Introduction

Model-based fault detection is based on generating a difference, known as a *residual*, between the predicted output value from the system model and the real output value measured by the sensors. If this residual is bigger than a threshold, then it is determined that there is a fault in the system. Otherwise, it is considered that the system is working properly. However, it is very important to analyse how the effect of model uncertainty is taken into account when determining the optimal threshold to be used in residual evaluation. In case that uncertainty is located in parameters (*interval model*), an interval observer has been shown to be a suitable strategy to generate such threshold. But, in general, computing an exact threshold using interval observers is time consuming because of the optimisation problem that must be solved at each time instant in order to avoid the problems presented in Stancu [15], namely: the wrapping effect, the interval function range evaluation and the uncertain parameter time dependency. The aim of this paper is to present a new algorithm for fault detection using interval observers, less computational demanding, based on a *forward/backward test*. Basically, this algorithm consists in two steps: first a *forward test* based on checking if measurements belong to the inner solution of the estimated output interval computed using an interval observation algorithm that only uses the vertices of the

C. Jermann et al. (Eds.): COCOS 2003, LNCS 3478, pp. 154–170, 2005.

parameter uncertain region and a backward test based on a consistency test between measurements and the interval model. Forward test cannot assure that a fault occurred when measurements are outside the inner solution because of its incompleteness. To check whether or not this measurement signals a fault, a consistency test must be performed to verify if there are system parameters that can explain this output value. This stage represents the **backward** test and is equivalent with a system identification using a single pair of input/output data.

The structure of rest of the paper is the following: in *Section 2*, fault detection based on interval observers is presented. In *Section 3*, forward and backward fault detection tests are introduced. In *Section 4*, the implementation of forward and backward tests is presented. Finally, in *Section 5*, the forward-backward fault detection algorithm is tested on the nozzle servosystem of a gas turbine.

2   Problem Formulation

2.1   Residual Generation and Robustness Issues

Considering a non-linear dynamic system in discrete-time with disturbances (or noises) $d(k)$, faults $f(k)$ and the modeling uncertainty located in parameters θ that affect the behaviour of the system, the state-space relationship can be written as

$$x(k+1)= g(x(k),u(k),d(k),f(k),\theta)$$
$$y(k)= h(x(k),u(k),d(k),f(k),\theta) \tag{1}$$

where:

- $x \in \Re^{nx}$, $u \in \Re^{nu}$ and $y \in \Re^{ny}$ are state, input and output vectors of dimension nx, nu and ny respectively;
- $d \in \Re^{nd}$, $n \in \Re^{nn}$ and $f \in \Re^{nf}$ are process disturbances, measurement noise and faults of dimension nd, nn and nf respectively;
- g and h are the state space and measurement non-linear function;
- θ is the vector of uncertain parameters of dimension p with their values bounded by a compact set $\theta \in \Theta$ of box type, i.e., $\Theta = \{\theta \in \Re^{p} / \underline{\theta} \le \theta \le \overline{\theta}\}$. This type of model is known as an **interval model**.

Model-based fault detection algorithms generally consist of two stages [4]:

- **Residual generation**: The model and the input/output measurements are used to determine residuals, which describe the degree of consistency between the plant and the model behaviour.
- **Residual evaluation**: The residual is evaluated in order to detect and isolate faults.

A **residual generator** can be constructed by

$$r(k)= y(k)-\hat{y}(k) \tag{2}$$

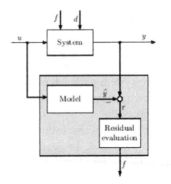

Fig. 1. Model based fault detection

where: $r(k)$ is the vector of residuals, $y(k)$ and $\hat{y}(k)$ are vectors of measured and estimated outputs. Ideally, the residuals should only be affected by the faults, therefore when a residual deviates from zero a fault should be indicated. However, the presence of disturbances, noise and modeling errors causes the residuals to become nonzero interfering with the detection of faults. Therefore, the fault detection procedure must be **robust** in the face of these undesired effects. Robustness can be achieved in the residual generation (**active robustness**) or in the decision making stage (**passive robustness**) [2]. The passive approach is based not in avoiding the effect of uncertainty in the residual, but in propagating the effect of uncertainty to the residual. If the residual

$$r(k) = y(k) - \hat{y}(k) \in \left[\underline{r}(k), \bar{r}(k)\right] \qquad (3)$$

no fault can be indicated, because the residual value can be due to the parameter uncertainty.

2.2   Passive Robustness Based on Interval Observers

Instead of using directly the interval model of the monitored system to produce the output estimation, an observer will be considered. Considering a non-linear interval model, the **interval observer** equation with a Luenberger-like structure without noise, faults and disturbances is:

$$\hat{x}(k+1) = g(\hat{x}(k), u(k), \theta) + K(y(k) - \hat{y}(k))$$
$$\hat{y}(k) = h(\hat{x}(k), u(k), \theta) \qquad (4)$$

where:

- $\hat{x} \in \Re^{nx}$ and $\hat{y} \in \Re^{ny}$ are estimated state and output vectors of dimension nx and ny respectively;
- K is the gain of the observer designed to guarantee observer stability for all $\theta \in \Theta$.

The evaluation of the interval for estimated output provided by the interval observer (3): $\underline{y}(k), \overline{y}(k)$ in order to evaluate the interval for residuals: $\underline{r}(k), \overline{r}(k)$ will be computed by means of a ***worst-case (or interval) observation***. It consists in computing a region of confidence for system state set $\hat{X}_{k+1}$, based on the confidence region for the system parameters Θ, the previous confidence region for the system state set $\hat{X}_k$ (in the case of one step algorithms), or the previous confidence regions for the system state set $\hat{X}_k, \ldots, \hat{X}_{k-L}$ (in the case of sliding time window algorithms) and the measurements available.

The observer equation (4) can be reorganised as a system with one output and two inputs, according to

$$\hat{x}(k+1) = g_o(\hat{x}(k), u_o(k), !)$$
$$\hat{y}(k) = h(\hat{x}(k)) \tag{5}$$

where: $u_o(k) = [u(k) \quad y(k)]^t$ and
$g_o(\hat{x}(k), u_o(k), !) = g(\hat{x}(k), u(k), !) + Ky(k) - Kh(\hat{x}(k), u(k), !)$ is the observer non-linear function. Then, worst-case observation can be formulated as a ***worst-case simulation.***

3   Forward and Backward Tests in Fault Detection

Because of the problems that can appear in interval observation and the complexity and computational exponential time for algorithms, passive robust fault detection for the interval non-linear models is far from a straightforward problem as it was shown in [15].

In this paper we propose an alternative way to deal with the passive robust fault detection based on interval models. A new fault detection algorithm that combine approximate interval observation, that can be viewed as a direct interval mapping (forward test), with the use of the inverse interval mapping (backward test) using interval constraint satisfaction algorithms is proposed. In the ***forward test***, a direct mapping based on an interval observer is used to propagate from step to step the interval of the possible system outputs, and then checking if the measurement coming from sensors belongs or not to such interval. On the other band, in the ***backward test***, an inverse mapping, also based on an interval observer, is used to check if the measurement invalidates or not the interval model used to monitor the system.

3.1   Forward Test for Fault Detection

Model based fault detection using observers is based on estimating each system output using measured inputs and outputs and the model. In fact, an observer can be viewed as a multi input single output (MISO) system according to (5). Then, considering the interval of uncertain parameters Θ, and assuming that g_o^k is the observer function that transport the system from initial state to the present state, the forward interval propagation of Θ will produce an interval hull for system states $\Box X_k$ or for system outputs $\Box Y_k$ (considering in this case that $y_k = h(x_k)$) such that at time k will provide the following fault detection test:

$$y_{measured}(k) \notin \Box Y_k \to \text{fault} \tag{6}$$
$$y_{measured}(k) \in \Box Y_k \to \text{no fault can be indicated} \tag{7}$$

However, in practice the interval hull for system output $\Box Y_k$ is very hard to compute [13]. On the other hand, inner and outer approximations to this interval hull can be computed. An inner approximation of $\Box Y_k$, denoted by $\Box \check{Y}_k$, can be computed by the vertex algorithm [8]. While an outer solution of $\Box Y_k$, denoted by $\Box \widehat{Y}_k$, can be computed for example using the optimisation algorithm presented in [14]. The use of these approximate solutions of $\Box Y_k$ will provide two different set of tests for fault detection:

Table 1. Fault detection based on inner and outer solutions

Outer solution fault detection test	Inner solution fault detection test
$y_{measured}(k) \notin \Box \widehat{Y}_k \to \text{fault}$	$y_{measured}(k) \notin \Box \check{Y}_k \to \text{undetermined}$
$y_{measured}(k) \in \Box \widehat{Y}_k \to \text{undetermined}$	$y_{measured}(k) \in \Box \check{Y}_k \to \text{no fault}$

These two sets of tests are complementary. The outer solution test allows to detect the faulty situations while the inner solution the non-faulty. There is an undecided zone corresponding to the following situation:

$$y_{measured}(k) \in \Box \widehat{Y}_k \quad \text{but} \quad y_{measured}(k) \notin \Box \check{Y}_k \tag{8}$$

This region can only be reduced refining either the inner or the outer approximations of $\Box Y_k$, but always at a high cost since some bisection mechanism should be introduced. A fault detection algorithm that combine this inner and outer forward tests is proposed by [1].

However, in case the forward test would be applied to an observer with multiple outputs (MIMO system), it would not work correctly in general. The reason is because the forward test represents the output space using intervals, but in general, this output space in the case of several outputs is not an interval but instead a more complex region (in general non-convex if the observer is non-linear) since there are dependencies between outputs and parameters. The phenomenon is intuitively illustrated in Figure 2 where it can be seen the spurious outputs included because the real region $A'B'C'D'$ is wrapped using a box. Then, for example, point S will belong to the output envelopes in time domain. These spurious outputs corresponds to parameters in the white zone added to the original parameter presented in the Figure 2, i.e., the original parameter uncertain domain has been "artificially" augmented. Then, point S

can be obtained with a combination between one point from the initial states domain *ABCD* and one point from "augmented" white zone in the parameter uncertainty domain. If point *S* belongs to the system output envelopes, the decision that point *S* corresponds to a normal behaviour is not correct in this case since it has been produced by a set of parameters out of the original parameter domain. As conclusion, in this paper the focus of the proposed algorithms will be on the case of multiple input and single output (MISO) observers where the output space can be correctly represented with an interval, leaving for further research the case of MIMO observers.

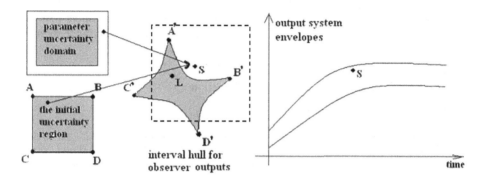

Fig. 2. Inclusion of spurious outputs in the case of a multiple-output observer

3.2 Backward Test for Fault Detection

In this paper the fault detection test based on the combined used of inner and outer forward tests will be improved in the following way. Since outer solutions of $\Box Y_k$ solving the interval observation problem are generally hard to obtain [15] instead the following backward test is proposed to detect the faulty situations assuming zero initial conditions[1]:

$$\exists \theta \in \Theta \text{ such that } y_{measured}(k) = h\left(g_o^k(\theta)\right) \tag{9}$$

where Θ is the interval of uncertain parameters and g_o^k is the observer function that transports the system from initial state to the present state, and h is the measurement function. In case that such test is not verified a fault can been indicated, otherwise no fault can be indicated. Additionally, the backward test allows very easyly the inclusion of additive bounded noise $\left[y_{measured}(k) - \varepsilon, y_{measured}(k) + \varepsilon\right]$ being ε the noise bound.

---

[1] In case that initial conditions are not zero can be easily included just modifying (9) adding $\exists x_o \in X_o$ and considering the dependence of $h\left(g_o^k(\theta)\right)$ on the initial condition.

Alternatively, test (9) can be viewed as computing the set of parameters $"\,"_{consistent}$ consistent with the measured output $y_{measured}$. Then, the backward fault detection can be stated as

$$"\cap"_{consistent} = \varnothing \rightarrow \text{fault} \tag{10}$$

$$"\cap"_{consistent} \neq \varnothing \rightarrow \text{no fault can be indicated} \tag{11}$$

The result $"\cap"_{consistent} \neq \varnothing$ in practice will result in an undecided test since it will be efficiently implemented using interval constraint satisfaction algorithms that use only local consistency and do not use bisections and provide only and outer solution for $"\cap"_{consistent}$, as it will be explained later (*see Section 4*).

3.3 Forward-Backward Tests for Fault Detection

In the forward fault detection tests presented in Table 1:

- the test based on an inner solution is used to check the consistency between a measurement and the interval model, testing if the measurement belongs or not to the predicted inner interval. In the case of passing the test no fault can be indicated, otherwise nothing can be stated,
- while the test based on an outer solution is used to detect the fault occurrence when a measurement does not belong to the predicted outer interval since the interval model is invalidated. Otherwise, nothing can be stated.

As it was presented in [15], in general ,it is very difficult to compute an outer solution for interval observation and therefore to prove that a measurement invalidates the interval model. In order to avoid this hard computational problem, here the forward test based on the outer solution will be substituted with a backward test based on interval constraint satisfaction that will allow to detect a fault when a measurement invalidates the interval model. It is known that the constraint propagation approach is a very powerful tool to proof the no-consistency. With such modification the fault detection strategy presented in Table 1, now it will be composed by two tests presented in Table 2. This fault detection strategy will be called in the following as *forward-backward*.

Table 2. Fault detection based on forward and backward tests

Backward fault detection test	Forward (inner) fault detection test
$"\cap"_{consistent} = \varnothing \rightarrow$ fault $\quad$ $"\cap"_{consistent} \neq \varnothing \rightarrow$ undeterrmined	$y_{measured}(k) \notin \Box \breve{Y}_k \rightarrow$ undetermined $\quad$ $y_{measured}(k) \in \Box \breve{Y}_k \rightarrow$ no fault

Using the forward-backward test, of course, there still will be an undecided zone corresponding to the following situation:

$$" \cap "_{consistent} \neq \emptyset \quad but \quad y_{measured}(k) \notin \Box \check{Y}_k \tag{12}$$

This region can only be reduced refining either the inner approximation of $\Box Y_k$ either the backward test that provides $"_{consistent}$, but always at a high cost since some bisection mechanism should be introduced.

4 Implementation of Forward-Backward Algorithm

Because of interval observation computational complexity associated to the computation of the exact output interval, the forward-backward algorithm has a practical significance when applied to detect system faults on-line where real-time performance is required. The aim of this algorithm is not computing the exact interval for estimated measurements but instead on verifying if they are consistent with real measurements.

This algorithm is based on a two decision tests (Table 2). First test checks if real measurements are inside to inner approximation of the interval for estimated measurements (forward) using a computational cheap algorithm. If a measurement belongs to the inner solution, the measurement does not invalidate the interval model. Second test is activated when measurements are outside the inner approximation of the interval for estimated measurements (backward). In this case, the measurement is used to invalidate the interval model detecting the fault in case of invalidation is confirmed. This test guarantees that any fault that invalidates the interval model is detected.

4.1 Implementation of the Forward Test

The forward test requires an inner solution of the interval for estimated system outputs. Kolev's algorithm [8] based on vertex simulation will produce an inner solution, i.e. a subset of solutions when the interval system is non-monotonic respect all the states. The inner solution provided by Kolev's algorithm coincides with the exact interval hull of the solution set for some particular systems, in particular, in the case of systems without the wrapping effect, according to [10]. Those systems satisfy the isotonic property according to [3]. And, moreover, according to [8], for a constant input $u(k)=u$, the inner solution coincides over the time intervals $[0,k_1]$ and $[k_2,\infty)$ with the exact solution.

Kolev's algorithm provides an ***inner solution*** for the interval observation problem by determining the interval vector $! \check{Y}_k = \left[\underline{\check{y}}(k), \overline{\check{y}}(k) \right]$ through the solution of the following global optimisation problems:

$$\overline{\check{y}}(k) = max \ \boldsymbol{h}(\boldsymbol{g}_o^k(!)) \quad and \quad \underline{\check{y}}(t) = min \ \boldsymbol{h}(\boldsymbol{g}_o^k(!)) \tag{13}$$

subject to: $! \in V(")$ where $h(g_o^k)$ denotes a solution of the output estimated trajectory of interval observer (5) at time k for some value of the vector of parameters in
$V(")$ that denotes the set of vertices of the uncertain parameter set $"$. The interval
vector $! \; \breve{Y}_k$ provides an inner solution of $! \; Y_k = \left[\underline{y}(k), \overline{y}(k)\right]$ for time k since

$$\breve{\underline{y}}(k) \geq \underline{y}(k) \tag{14}$$

$$\overline{\breve{y}}(k) \leq \overline{y}(k) \tag{15}$$

because only a subset (the vertices) of the parameter set $"$ are considered.

4.2 Implementation of Backward Test

The backward test can be viewed as the computation of the inverse image of
measured   output $y_{measured}(k)$,   that   it   is   known   to   belong   to
$\left[y_{measured}(k) - \varepsilon, y_{measured}(k) + \varepsilon\right]$ assuming that the noise is bounded by ε, through
the observer output estimated trajectory providing the set of parameters consistent
with it

$$"_{consistent} = \left(h\left(g_o^k\right)\right)^{-1}(y_{measured}) \tag{16}$$

Once the set of parameters consistent with the measurement is obtained, the fault
detection test is given by (10) and (11).

Jaulin in [7] has proposed an algorithm called SIVIA that computes the inverse
image of an interval function using subpavings. However, when the dimension of the
set to characterize is of high dimension since SIVIA uses bisection in all directions
the computational complexity explodes. In this case the use of contractors and bisection when needed using constraint satisfaction principles (constraint projection) save a
lot of computation.

An ***interval constraint satisfaction problem*** (ICSP) can be formulated as a 3-tuple
$\mathcal{H} = (\mathcal{V}, \mathcal{D}, \mathcal{C})$, where $\mathcal{V} = \{v_1, \cdots, v_n\}$ is a finite set of variables, $\mathcal{D} = \{[v_1], \cdots, [v_n]\}$
is the set of their domains represented by closed real intervals and $\mathcal{C} = \{c_1, \cdots, c_n\}$ is a
finite set of constraints relating variables of $\mathcal{V}$. A ***point solution*** of $\mathcal{H}$ is a n-tuple
$(\tilde{v}_1, \cdots, \tilde{v}_n) \in \mathcal{V}$ such that all constraints $\mathcal{C}$ are satisfied. The set of all point solutions
of $\mathcal{H}$ is denoted by $S(\mathcal{H})$. This set is called the ***global solution set***. The variable $v_i \in \mathcal{V}_i$
is ***consistent*** in $\mathcal{H}$ if and only if:

$$\forall v_i \in \mathcal{V}_i \; \exists (\tilde{v}_1 \in [v_1], \cdots, \tilde{v}_i \in [v_i], \cdots, \tilde{v}_n \in [v_n]) \mid$$
$$(\tilde{v}_1, \cdots, \tilde{v}_n) \in S(\mathcal{H}) \tag{17}$$

The solution of an ICSP is said to be ***globally consistent***, if and only if every variable
is consistent. A variable is ***locally consistent*** if and only if it is consistent with respect
to all directly connected constraints. Thus, the solution of an ICSP is said to be locally
consistent if all variables are locally consistent. Several algorithms can be used to

solve this type of problem, including Waltz's *local filtering* algorithm [17] and Hyvönen's *tolerance propagation* algorithm [5]. The first only ensures locally consistent solutions while the second can guarantee global consistent solutions.

The principle of algorithms for solving ICSP using local consistency techniques consists essentially in iterating two main operations, *domain contraction* and *propagation*, until reaching a stable state. Roughly speaking, if the domain of a variable v_i is locally contracted with respect to a constraint c_j, then this domain modification is propagated to all the constraints in which v_i occurs, leading to the contraction of other variable domains and so on. Then, the final goal of such strategy is to contract as much as possible the domains of the variables without loosing any solution by removing inconsistent values through the *projection* of all constraints. To project a constraint with respect to some of its variables consists in computing the smallest interval that contains only consistent values applying a contraction operator.

Being incomplete by nature, these methods have to be combined with enumeration techniques, for example bisection, to separate the solutions when it is possible. Domain contraction relies on the notion of *contraction operators* computing over approximate domains over the real numbers.

According to (9), the backward test can be formulated as an interval constraint satisfaction problem assuming again zero initial conditions

$$
\begin{aligned}
&! \in '', \varepsilon \in \left[\underline{\varepsilon}, \overline{\varepsilon}\right] \\
&y_{measured}(k) \in \left[y_{measured}(k) - \varepsilon, y_{measured}(k) + \varepsilon\right] \\
&y_{measured}(k) = h\!\left(g_o^k(!)\right)
\end{aligned}
\tag{18}
$$

The function $h(g_o^k)$ that denotes the output estimated trajectory from the initial condition is growing with time. In case of stable interval observer, it can be approximated using a time window L, $h(g_o^{k-L})$ instead of solving it with respect to the initial state. These modifications reduce the computation time, allowing operation in real time. The length of L and its relation with the approximation degree introduced using this approach has been studied by [12] in case of linear observers.

The solution of the above ICSP, if there exist, will only provide an outer approximation of $'' \cap ''_{consistent}$ denoting the set of parameters consistent with the measurement interval that belong to initial parameter set since local consistency is used. Therefore the fault detection test is undetermined when such outer approximation is not empty. On the other hand, if there is no solution, it means that there are no system parameter consistent with the measurement coming from the sensor and it can assured that a fault has occurred. The backward test based on ICSP can only be refined using bisections until an inconsistency is detection, or performing the consistency using more measurements taken at different time instances.

4.3   The Forward-Backward Fault Detection Algorithm

Finally, the proposed forward-backward algorithm fault detection algorithm can be formulated as:

Algorithm 1.* *Forward-backward fault detection algorithm

a) Forward fault detection test (Vertex Simulation)

The interval for the estimated system outputs $\square Y_k$ *at time instant k using the interval observer formulated as in (5) is obtained using vertex simulation (see Section 4.1). Let* $y_{measured}(k)$ *be the measurement coming from the sensor at time instant k. Then:*

- *If* $y_{measured}(k) \in \square Y_k \rightarrow$ **no fault** can be indicated, *i.e., the interval model is* **not invalidated** *by the measurement*
- *If* $y_{measured}(k) \notin \square Y_k$ *then GOTO b)*

b) Backward test (Parameter Consistency Check)

An outer approximation of $" \cap "_{consistent}$ *is obtained solving ICSP (18) for each measurement (see Section 4.2). Then:*
- *If* $" \cap "_{consistent} = \varnothing$ *then* **model invalidated** *by* $y_{measured}(k)$ *and then a* **fault** *can be indicated.*
- *If* $" \cap "_{consistent} \neq \varnothing$ *the test is* **undecided**, *being necessary more measurements to make a decision.*

---

5   Application

The forward-backward fault detection algorithm will be tested using a real benchmark problem, pinpointing its advantages and drawbacks.

The example is based on the nozzle servosystem of a gas turbine and comes from the TIGER ESPRIT project [16]. The goal of TIGER project was to monitor a complex dynamic system such a gas turbine in real-time and make an assessment of whether it is working properly.

5.1   Interval Observer and Its Properties

Mathematical equations that describe the behaviour of observer for the nozzle servosystem are

$$x_1(k) = a_1 x_1 (k-1) + a_2 (r (k)-x_2 (k)) + a_3(r (k-1)-x_2 (k-1)) + a_4 \qquad (19)$$
$$x_2 (k) = x_2 (k-1) + \theta_1 x_1 (k-1) + \theta_2$$

where: r is the set-point for the nozzle position in degrees , x_1 is the input of the actuator that positions the nozzles, x_2 is the nozzle position in degrees and the system parameters were obtained from experimental data using parameter identification techniques being:

$$a_1 = 0.7572 \quad a_2 = 0.2298 \quad a_3 = 0.1202 \quad a_4 = 0.0202$$

with uncertain parameters:

$$\theta_1 \in [0.1032, 0.3372] \qquad \theta_2 \in [0.0155, 0.0923] \qquad (20)$$

Making some algebraically arrangements, the system matrix will be:

$$A = \begin{bmatrix} 0.7572 - 0.22980\theta_1 & -0.35 \\ \theta_1 & 1 \end{bmatrix} \qquad (21)$$

This system suffers from the wrapping effect because do not fulfil the propriety of isotony[2] (Cugueró, 2002). It can also be seen that the system suffers an instable wrapping effect because the system matrix is not contractive[3], since $\|A(\theta_1)\|_\infty \leq 1.3372$.

Uncertain parameters in (20) are considered unknown in their intervals but time-invariant (Puig, 2003a), i.e., θ_1 and θ_2 can be considered as extended states with the following dynamics:

$$\theta_1(k) = \theta_1(k-1) \qquad (23)$$
$$\theta_2(k) = \theta_2(k-1) \qquad (24)$$

Then, the interval observer (19) is non-linear because of the products between parameters and system states.

5.2 Inner and Outer Approximation of the Exact Output Estimated Interval

In the Figure 3, the exact and the inner approximation of the estimated output interval are presented for a step input and initial conditions $x_1(0)=-4.5$ and $x_2(0)=-0.42$. Note that the exact solution is obtained using GlobSol Solver [7] as a global optimiser with a precision 10^{-5} on a Pentium 475 MHz with a very high computational time [13]. The inner solution is obtained using Kolev's algorithm. Because of the incompleteness of the method that generates the inner solution, it does not coincide with the exact solution at any time instant. In this case because a constant input is introduced, the inner solution coincides over the time intervals $k \in [0,9]$ and $k \in [38,+\infty)$ with the exact solution, being consistent with results presented in [8].

---

[2] A non-linear system has the isotony property iff the variation of the state function respect all the states and parameters is positive.

[3] A non-linear function $f : R^n \to R^n$ is a contraction mapping means that if there is a number s, with $0<s<1$, so that for any vectors x and y we must have $d(f(x), f(y)) \leq sd(x,y)$ where s is the contractivity. As special in case of linear systems $\|A\|_\infty = s$.

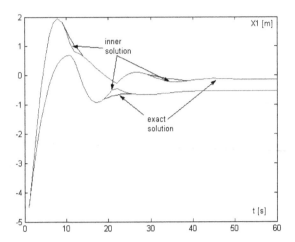

Fig. 3. Exact and inner approximation of the estimated output interval

In the Figure 4, the exact, inner and various outer solutions with different degree of approximation of the estimated output interval are presented. The outer approxima-tions of the estimated output interval are obtained solving the following consistency problem considering as it has been stated in (23) and (24) that uncertain parameters are time-invariant:

$$\hat{y}(k) = x_1(k) = A(\theta_1, \theta_2)x(0) + \sum_{j=0}^{k-1} A^{k-1-j}(\theta_1, \theta_2)B(\theta_1, \theta_2)$$

$\theta_1 \in [0.1032,\ 0.3372]$

$\theta_2 \in [0.0155,\ 0.0923]$

$y(k) \in (-\infty, +\infty)$

where:

$$A = \begin{vmatrix} 0.7572 - 0.2298\theta_1 & -0.35 \\ \theta_1 & 1 \end{vmatrix} \ \text{and} \ B = \begin{vmatrix} -0.2248 - 0.2298\theta_2 \\ \theta_2 \end{vmatrix}.$$

using the Proj2D Solver [9] selecting different values for precision parameter ε. This precision parameter measures the number of bisections. Decreasing the precision parameter, the consistency problem solution tends to be global consistent, however the computation time increases a lot.

In order to obtain a non divergent outer solution the precision parameter must be decreased as it can be observed in Figure 4. For outer solution computation, the computational time is very big. For instance, selecting as the value for the precision degree $\varepsilon = 0.02$, the computational time is 180 s at time instant t=10 s, 297 s at time instant t=11 s, and 798 s at time instant t=13 s. These results allow to show the high computational complexity required to compute a non-divergent outer solution as it has been stated in *Section 3.1*. This is the main motivation to introduce an alternative way to obtain the same fault detection results than the ones provided by the outer forward test.

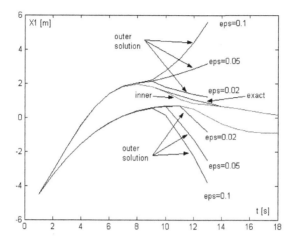

Fig. 4. Exact, inner and outer approximations of the estimated output interval with different degree of precision

5.3   Forward-Backward Test

In the Figure 5 (a), it is illustrated the situation when a measurement falls outside the inner solution. Because of its incompleteness, the backward test is activated, as a complementary test, in order to check the consistency between the measurement and the output interval estimation provided by the interval observer. In case of a solution is found (Figure 5(b)), the algorithm will be in the undecided situation according to *Algorithm 1*, because an outer approximation of $"\cap"_{consistent}$ is obtained by solving the ICSP (18).

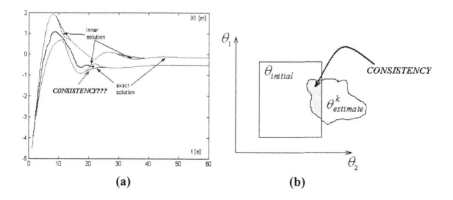

Fig. 5. (a) Forward and **(b)** backward test in case that backward test is undetermined

In the next time step, the measurement continues to be outside the inner solution, then again the backward test is activated (Figure 6(a)). However, in this case according to Figure 6(b) the backward test will provide an inconsistency, i.e., the measurement invalidates the interval model. Then, a fault occurrence can be assured. From Figure 6(a), it can be observed that at this time step the measurement also is out of the exact solution and it would be detected if it could be computed.

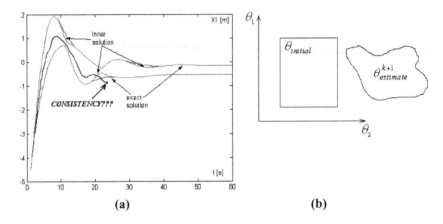

(a)   (b)

Fig. 6. Forward and backward test in case that backward test is determined

As it can be observed from this case study, the forward-backward algorithm is a powerful tool for fault detection. The advantage of this algorithm is that the very hard computational and possible more conservative outer solution is not needed. Only a cheaper computational inner solution computed on line with the real process is required. When a measurement falls outside the inner solution, the backward test will be activated. This test is not so computationally intensive as the outer approximation is, because it makes uses of local consistency algorithms based on contractors avoiding the use of bisections. However, there is still an undecided zone bigger than the distance between the inner and exact solution. This undecided zone exceeds the exact solution because of the local consistency used. It can only be reduced using bisections in the backward test or more measurements taken at different time instants.

6 Conclusions

Considering the problems that appear in interval observation using regions or real trajectories [15], a new algorithm for fault detection is proposed. This algorithm uses a vertex simulation (forward test) to compute an inner approximation of the estimated output interval because of its lower computational complexity. However, because of the incompleteness of such test, a backward test based on interval constraint satisfaction is used when a measurement coming from the sensor falls outside the inner solution. When this measurement belongs to the region between the inner solution and the exact solution (unknown), the backward test solving the ICSP using local consistency provides an nonempty outer approximation of $"\cap"_{consistent}$ and we cannot decide that this measurement represents a faulty or a normal situation. However, when the measurement is outside the undecided zone, the consistency test provides very quickly that the outer approximation of $"\cap"_{consistent}$ is empty assuring that a fault occurred. Finally, this new fault detection algorithm has successfully been applied to detect faults in a nozzle servosystem of gas turbine.

In conclusion, this forward-backward algorithm is developed in order to be applied in fault detection applications where real-time operation is needed.

As a future work we want to minimise as much is possible this undecided zone, and to combine the forward-backward algorithm with another tests in order to decide about the measurements that belong to the undecided zone.

Acknowledgements

This paper is supported by CICYT (DPI2002-03500), by Research Commission of the "Generalitat de Catalunya" (group SAC ref.2001/SGR/00236) and by DAMADICS FP5 European Research Training Network (ref. ECC-TRN1-1999-00392).

References

1. Armengol, J., Vehí, J., Travé-Massuyès, L., Sainz, M.A. "Application of Modal Intervals to the Generation of Error-Bounded Envelopes". Reliable Computing 7(2): 171-185, February 2001.
2. Chen J. and R.J. Patton. "Robust Model-Based Fault Diagnosis for Dynamic Systems". Kluwer Academic Publishers. 1999.
3. Cugueró, P., Puig, V., Saludes, J., Escobet, T. "A Class of Uncertain Linear Interval Models for which a Set Based Robust Simulation can be Reduced to Few Pointwise Simulations". In Proceedings of Conference on Decision and Control 2002 (CDC'02). Las Vegas. USA. 2002.
4. Gertler, J.J. "Fault Detection and Diagnosis in Engineering Systems". Marcel Decker. 1998.
5. Hyvönen, E. "Constraint Reasoning based on Interval Arithmetic: The Tolerance Approach". Artificial Intelligence, 58 pp. 71-112, 1992.
6. Jaulin, L., M. Kieffer, O. Didrit and E. Walter. Applied Interval Analysis, with Examples in Parameter and State Estimation, Robust Control and Robotics. Springer-Verlag. London. 2001.
7. Kearfott, R.B. "Rigorous Global Search: Continuous Problems". Kluwer Academic Publishers. Dordrecht. 1996.
8. Kolev, L.V. "Interval Methods for Circuit Analysis". Singapore. World Scientific. 1993.
9. Dao, M., Jaulin, L. "Proj2D Solver". http://www.istia.univ-angers.fr/~dao/. 2003.
10. Nickel, K.. "How to fight the wrapping effect". In K. Nickel ed. "Interval Analysis 1985". Lecture Notes in Computer Science, No. 212, pp. 121-132. Springer-Verlag. 1985.
11. Puig, V., Quevedo, J., Escobet, T., De las Heras, S. "Robust Fault Detection Approaches using Interval Models". IFAC World Congress (b'02). Barcelona. Spain. 2002.
12. Puig, V., Quevedo, J., Escobet, T., Stancu, A. "Passive Robust Fault Detection using Linear Interval Observers". IFAC Safe Process, 2003. Washington. USA. 2003.

13. Puig, V., Saludes, J., Quevedo, J. "Worst-Case Simulation of Discrete Linear Time-Invariant Dynamic Systems", Reliable Computing 9(4): 251-290, August 2003.

14. Stancu, A., Puig, V., Quevedo, J., Patton R. J. "Passive Robust Fault Detection using Non-Linear Interval Observers: Application to the DAMADICS Benchmark Problem". IFAC Safe Process, 2003. Washington. USA.

15. Stancu, A., Puig, V., Cugueró, P., Quevedo, J. "Benchmarking on Approaches to Interval Observation Applied to Robust Fault Detection", 2nd International Workshop on Global Constrained Optimization and Constraint Satisfaction (Cocos '03), November 2003, Lausanne, Switzerland.

16. Travé-Massuyes, L. R. Milne. "TIGER: Gas Turbine condition monitoring using qualitative model based diagnosis". IEEE Expert Intelligent Systems and Applications, May-June. 1997.

17. Waltz, D. (1975). "Understanding line drawings of scenes with shadows". P.H. Winston Ed. "The Psychology of Computer Vision". McGraw-Hill, New York, pág 19-91.1975.

Benchmarking on Approaches to Interval Observation Applied to Robust Fault Detection

Alexandru Stancu, Vicenç Puig, Pep Cugueró,
and Joseba Quevedo

Automatic Control Department - Campus de Terrassa,
Universidad Politécnica de Cataluña (UPC),
Rambla Sant Nebridi, 10. 08222 Terrassa (Spain)
{Alexandru.Stancu, Vicenc.Puig, Josep.Cuguero,
Joseba.Quevedo}@upc.es

Abstract. Model-based fault detection is based on generating a difference, known as a residual, between the predicted output value from the system model and the real output value measured by the sensors. If this residual is bigger than a threshold, then it is determined that there is a fault in the system. Otherwise, it is considered that the system is working properly. However, it is very important to analyse how the effect of model uncertainty is taken into account when determining the optimal threshold to be used in residual evaluation. In case that uncertainty is located in parameters (interval model), an interval observer has been shown to be a suitable strategy to generate such threshold. However, interval observers can present several problems that in order to be solved, existing approaches require computational demanding algorithms. The aim of this paper is to study the viability of using region based approaches coming from the interval analysis community to solve the interval observation problem. Region based approaches are appealing because of its low computational complexity but they suffer from the wrapping effect. On the other hand, trajectory based approaches are immune to this problem but their computational complexity is higher. In this paper, these two interval observation philosophies will be presented, analysed and compared using in two examples.

1 Introduction

Fault detection methods based on the mathematical model of the system use the difference between the predicted value from the model and the real value measured by the sensors to detect faults. This difference known as *residual* will be compared with a threshold value. If the residual is bigger than the threshold, then it is determined that there is a fault in the system. Otherwise, it is considered that the system is working properly. However, when modelling a physical dynamic system with a mathematical model, there is always some uncertainty that will interfere in the detection process. In the case of uncertainty in the parameters, a model whose parameter values are bounded by intervals, known as an *interval model*, is usually considered. The *robustness* of a fault detection system means that it must be only sensitive to faults, even in the presence of model-reality differences [2]. Robustness can be achieved at residual

C. Jermann et al. (Eds.): COCOS 2003, LNCS 3478, pp. 171–191, 2005.
© Springer-Verlag Berlin Heidelberg 2005

generation or evaluation phase. Most of the robust residual evaluation methods are based on an adaptive threshold changing in time according to the plant input signal and taking into account model uncertainty. These last years the research of adaptive thresholding algorithms that use interval models for fault detection has been a very active research area [20]. In [22] interval observers applied to robust fault detection have been introduced and an algorithm based on optimisation based interval simulation is proposed [23]. However, the computational complexity of this approach is high, so less computational demanding algorithms should be devised. This is the aim of this paper. To achieve this goal, region based approaches coming from the interval analysis community will be analysed since their low computational complexity. However, they can suffer from the wrapping effect.

The structure of the rest of the paper is the following: in *Section 2*, fault detection based on interval observers is presented. In *Section 3*, problems associated to interval observation are introduced. In *Section 4*, region based approaches are presented while in *Section 5* trajectory based approaches are considered. Finally, in *Section 6*, two test examples will be used to compare their performance with trajectory based approaches that are immune to this problem but whose computational complexity is higher.

2   Robust Fault Detection

2.1   Residual Generation and Robustness Issues

Considering a non-linear dynamic system in discrete-time with disturbances (or noises) $d(k)$,   faults $f(k)$ and the modeling uncertainty located in parameters $\boldsymbol{\theta}$ that affect the behaviour of the system, the state-space relationship can be written as

$$x(k+1) = g(x(k), u(k), d(k), f(k), \boldsymbol{\theta})$$
$$y(k) = h(x(k), u(k), d(k), f(k), \boldsymbol{\theta})$$

(1)

where:

- $x \in \Re^{nx}$, $u \in \Re^{nu}$ and $y \in \Re^{ny}$ are state, input and output vectors of dimension nx, nu and ny respectively;
- $d \in \Re^{nd}$, $n \in \Re^{nn}$ and $f \in \Re^{nf}$ are process disturbances, measurement noise and faults of dimension nd, nn and nf respectively;
- g and h are the state space and measurement non-linear function ;
- $\boldsymbol{\theta}$ is the vector of uncertain parameters of dimension p with their values bounded by a compact set $\boldsymbol{\theta} \in \boldsymbol{\Theta}$ of box type, i.e., $\boldsymbol{\Theta} = \{\theta \in \Re^{p} \mid \underline{\theta} \leq \theta \leq \overline{\theta}\}$. This type of model is known as an ***interval model***.

Model-based fault detection algorithms generally consist of two stages [2]:

- ***Residual generation***: The model and the input/output measurements are used to determine residuals, which describe the degree of consistency between the plant and the model behaviour.

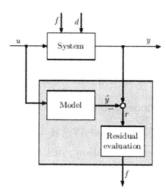

Fig. 1. Model based fault detection

- *Residual evaluation*: The residual is evaluated in order to detect and isolate faults.

A *residual generator* can be constructed by

$$r(k) = y(k) - \hat{y}(k)$$ (2)

where: $r(k)$ is the vector of residuals, $y(k)$ and $\hat{y}(k)$ are vectors of real and estimated measurements. Ideally, the residuals should only be affected by the faults. However, the presence of disturbances, noise and modeling errors causes the residuals to become nonzero interfering with the detection of faults. Therefore, the fault detection procedure must be *robust* in the face of these undesired effects. Robustness can be achieved in the residual generation (*active robustness*) or in the decision making stage (*passive robustness*) [2]. The passive approach is based not in avoiding the effect of uncertainty in the residual, but in propagating the effect of uncertainty to the residual. Let $\left[\underline{\hat{y}}(k), \overline{\hat{y}}(k) \right]$ be the interval for predicted output using model (1) considering parameter model uncertainty, then no fault can be indicated while the residual satisfies

$$r(k) = y(k) - \hat{y}_c(k) \in \left[- \Delta\hat{y}(k), \Delta\hat{y}(k) \right] = \left[\underline{r}(k), \overline{r}(k) \right]$$ (3)

where: $\hat{y}_c(k) = \dfrac{1}{2}(\underline{\hat{y}}(k) + \overline{\hat{y}}(k))$ is the predicted output interval centre and $\Delta\hat{y}(k) = \dfrac{1}{2}(\overline{\hat{y}}(k) - \underline{\hat{y}}(k))$ its radius. Otherwise, a fault should be indicated.

Of course this approach has the drawback that faults that produce a residual deviation smaller than the residual uncertainty because of parameter uncertainty will be missed. Test (3) is equivalent to check if the measured output belongs to the interval of predicted outputs, i.e., to check if $y(k) \in [\hat{\underline{y}}(k), \overline{\hat{y}}(k)]$.

2.2 Passive Robustness Based on Interval Observers

Instead of using directly the model of the monitored system to estimate the interval of estimated outputs $[\hat{\underline{y}}(k), \overline{\hat{y}}(k)]$, an observer for this system will be considered.

A *non-linear interval observer* equation with a Luenberger-like structure for the system (1) can be introduced as a generalisation of a linear interval observer [22]:

$$\hat{x}(k+1) = g(\hat{x}(k), u(k), \theta) + K(y(k) - \hat{y}(k))$$
$$\hat{y}(k) = h(\hat{x}(k), u(k), \theta)$$
$$(4)$$

where:

- $\hat{x} \in \mathfrak{R}^{nx}$ and $\hat{y} \in \mathfrak{R}^{ny}$ are estimated state and output vectors of dimension nx and ny respectively;
- K is the gain of the observer designed to guarantee observer stability for all $\theta \in \Theta$.

The interval for estimated outputs provided by the interval observer (4), that will allow to evaluate the interval for residuals: $[\underline{r}(k), \overline{r}(k)]$, will be computed by means of an *interval* (or *worst-case*) *observation*. This consists in approximating at each time iteration the set of estimated system states $\hat{X}(k)$ and outputs $\hat{Y}(k)$ by its interval hull (the least interval box that contain this region), based on the set of uncertain parameters Θ, the previous approximations of the sets of estimated states $\hat{X}(k-1), ..., \hat{X}(0)$ and the measurements available $y(k-1)... y(0)$.

The observer equation (4) can be reorganised as a system with one output and two inputs, according to

$$\hat{x}(k+1) = g_o(\hat{x}(k), u_o(k), \theta)$$
$$\hat{y}(k) = h(\hat{x}(k), u(k), \theta)$$
$$(5)$$

where: $u_o(k) = [u(k)\quad y(k)]^t$ and
$g_o(\hat{x}(k), u_o(k), \theta) = g(\hat{x}(k), u(k), \theta) + Ky(k) - Kh(\hat{x}(k), u(k), \theta)$ is the observer non-linear function. Then, worst-case observation can be formulated as a *worst-case (or interval) simulation*. Existing algorithms can be classified according to if they compute the output interval using: one step-ahead iteration based on previous approximations of the set of estimated states (*region based approaches*), or a set of point-wise trajectories generated by selecting particular values of $\theta \in \Theta$ using heuristics or optimisation (*trajectory based approaches*). In this paper, these two groups of approaches will be compared.

3 Problems in Worst-Case Observation

Since the problem of worst-case observation can be reformulated as a problem of worst-case simulation, all the problems affecting worst-case simulation using intervals should be taken into account when dealing with worst-case observation [21]. These problems are described in the following.

3.1 The Wrapping Effect

The problem of wrapping is related to the use of a crude approximation (the interval hull) of the interval observer solution set and its iteration using one-step ahead recursion of the state space observer function. This problem does not appear if instead the estimated trajectory function $\hat{x}(k, u, y, \theta)$ is used. On the other hand, when using the one-step ahead recursion approach, at each iteration, the true solution set $\hat{X}(k)$ is wrapped into a superset feasible to construct and to represent the real region on a computer (in this paper, its interval hull $\square \hat{X}(k)$). Since the overestimation of the wrapped set is proportional to its radius, an spurious growth of the enclosures can result if the composition of wrapping and mapping is iterated [10]. This **wrapping effect** can be completely unrelated to the stability properties of the observer, and even stable observers are shown to exhibit exponentially fast growing enclosures that are useless for practical purposes. Not all the interval observers exhibit this problem. It has been shown that those that are monotone with respect to states do not present this problem. This kind of observers (systems) are known as **isotonic** [4] or **cooperative** [7].

3.2 The Interval Function Range Evaluation

Many approaches to interval observation need to evaluate the range of an interval function at each iteration in order to determine the interval for systems states. One possibility for evaluating the range of the function is to use interval arithmetic [12][13]. But, although the ranges of basic interval arithmetic operations are exactly the ranges of the corresponding real operations, this is not the case if the operations are composed. This phenomenon is termed as **interval dependence** or **multi-incidence problem** [12][13].

3.3 The Uncertain Parameter Time Dependency

An additional issue should be taken into account when an interval observer, as (4), is used: uncertain parameter time-invariance is not naturally preserved using one-step ahead recursion algorithms. If one-step recursion scheme is used, the set for system states $X(k+1)$ is approximated by a set computed using previous sets approximating system state region $X(k)$ and the set for uncertain parameters Θ. Then, the relation between parameters and states is not preserved since every parameter contained in the parameter uncertainty region Θ is combined with every state in the set approximating state region $X(k)$ when determining the new set approximating state region $X(k+1)$. Thus, recursive schemes based on one-step are intrinsically time varying. Time-invariance in parameters can only be guaranteed if the relation between parameters and states could be preserved at every iteration. One possibility to preserve this dependence is to derive a functional relation between states and parameters at every iteration that will transport the system from the initial state to the present state. Then, two approaches about the assumption of the time-variance of the uncertain parameters are possible:

- The *time-varying approach* which assumes that uncertain parameters are unknown but bounded in their confidence intervals and can vary at each time step [6][19].
- The *time-invariant approach* which assumes that uncertain parameters are unknown but bounded in their confidence intervals and they can not vary at each time step [8], [18].

4 Approaches Using Regions

All the algorithms described in this section produce only an outer (conservative) solution for worst-case observation. The propagation mechanism used in these algorithms produces an approximating region that includes all possible states in the exact solution region of estimated states based on previous approximating regions, but also includes spurious states. Therefore, the introduction of spurious states will inflate the uncertainty region, resulting in a superset of solutions that provides therefore an outer solution and producing in many cases an unstable simulation/observation. In this section algorithms that propagate regions developed by Moore [12][13], Lohner [11], Neumaier [15] and Kühn [10] will be presented and analysed regarding the problems presented in *Section 4*.

4.1 Moore's Algorithm [12][13]

The *absolute Moore's algorithm* is based on computing and propagating the interval hull of set of possible estimated states $\hat{X}(k)$, i.e., the smallest interval vector containing it:

$$\Box\hat{X}(k) = \left[\underline{\hat{x}}(k), \overline{\hat{x}}(k)\right] \tag{6}$$

where $\Box$ is used to denote the *interval hull* of $\hat{X}(k)$ and it can be computed determining for each component $\hat{x}_i(k)$ the maximum and the minimum according to

$$\overline{\hat{x}}_i(k) = max\{\hat{x}_i(k) : \hat{x}(k) \in \hat{X}(k)\}$$
$$\underline{\hat{x}}_i(k) = min\{\hat{x}_i(k) : \hat{x}(k) \in \hat{X}(k)\} \tag{7}$$

When the wrapping effect is present, the absolute Moore's algorithm diverges very quickly. In order to improve the absolute algorithm, Moore has proposed a *relative algorithm* based on the *interval mean-value theorem*[1]. The advantage of the relative

---

[1] **Mean Value Theorem**: if f: $R^n \to R$ is continuously differentiable on $D \subseteq R^n$ and $[a] \subseteq D$, then, for any x and b ∈ [a], f(x)=f(b)+f'(ξ)(x-b) for some ξ∈ [a].

algorithm at reducing the wrapping effect is that the region of system states is enclosed at each iteration in a moving co-ordinate system that matches the solution set. Associated with the set $\square\hat{X}(k)$ is the *central estimate* $\hat{x}_c(k)$ defined as follows:

$$\hat{x}_c(k)=\frac{1}{2}(\underline{x}(k)+\overline{x}(k)) \tag{8}$$

Then, state equations of interval observer (4), formulated as (5) can be linearised about the central estimate $\hat{x}_c(k)$ of $\square\hat{X}(k)$ as:

$$\begin{aligned} x(k+1,\boldsymbol{\theta}) &\cong g_0(\hat{x}_c(k),u_0(k),\boldsymbol{\theta}_c) \\ &+\frac{\partial g_0(x,u_0,\boldsymbol{\theta})}{\partial x}\bigg|_{x=\hat{x}_c(k)}(x(k)-\hat{x}_c(k)) \end{aligned} \tag{9}$$

as in the *Extended Kalman Filter* (EKF) [24]. Introducing:

$$A(x_c(k),\boldsymbol{\theta})=\frac{\partial g_0(x,u_0,\boldsymbol{\theta})}{\partial x}\bigg|_{x=\hat{x}_c(k)} \tag{10}$$

then (9) can be rewritten as:

$$\begin{aligned} x(k+1,\boldsymbol{\theta}) &\cong g_0(\hat{x}_c(k),u_0(k),\boldsymbol{\theta}_c) \\ &+A(\hat{x}_c(k),\boldsymbol{\theta})(x(k)-\hat{x}_c(k)) \end{aligned} \tag{11}$$

The relative Moore's algorithm is presented in the following.

Algorithm 1. Relative Moore algorithm

Assuming that $x(0)\in X_0$:

-   *compute the central estimate* $\hat{x}_c(k)$ *of* $\square\hat{X}(k)$
-   *propagate* $\square\hat{X}(k)$ *using the interval **mean-value theorem**:*

$$\square\hat{X}(k+1)\in g_0(\hat{x}_c(k),u(k))+A(\square\hat{X}(k))(\square\hat{X}(k)-\hat{x}_c(k)) \tag{12}$$

where: $g_0(\hat{x}_c(k),u(k))$ *is* $g_0(\hat{x}(k),u_0(k),\boldsymbol{\theta})$ *computed in the linearisation point and* $A(\square\hat{X}(k))$ *represents the interval Jacobian function.*

However, this method still suffers from the wrapping effect for non isotonic and in some ill-conditioned systems, as for example, systems with eigenvalues with very different magnitudes [14].

Because the Moore's algorithm was developed only for state uncertainty, when the systems parameters are allowed to contain intervals too, these parameters can be considered as an additional time-invariant uncertainty states according to Puig [19].

4.2 Lohner's Algorithm [11]

In those cases where Moore's algorithm is ill-conditioned, the algorithm should be modified according to Lohner [11]:

$$\Box \hat{X}(k+1) = S(k+1)\Box Z(k+1)$$

$$\Box Z(k+1) = S^{-1}(k+1)A(\Box \hat{X}(k))S(k)\Box Z(k)$$
$$+ S^{-1}(k+1)g_0(\hat{x}_c(k),u(k)) \tag{13}$$

where $S(k)$ is determined using a **QR-factorisation** method according to:

$$\Box Z(0) = \Box \hat{X}(0) \quad \text{and} \quad S(0) = I$$
$$\hat{S}(k) = m(A(\Box \hat{X}(k)))S(k)$$
$$\hat{S}(k) = Q(k)R(k)$$
$$S(k+1) = Q(k) \tag{14}$$

It is advisable to apply a pivoting strategy prior to the QR-factorization by sorting the columns of $\hat{S}(k)$ appropriately. The columns of this matrix span a good approximation of the exact solution set according to Lohner [11].

In case of a system including the uncertainty in the parameters, it must be used again the extended system (parameters as time-invariant states) [19].

One explanation why this method is successful at reducing the wrapping effect is that the region of system states is enclosed at each iteration in a moving orthogonal co-ordinate system that matches the solution set. Lohner has proposed the orthogonal transformation in order to obtain the matrix product $S^{-1}(k+1)A(\hat{X}(k))S(k)$ upper triangular and in this case the condition for avoiding the wrapping effect $\rho(A) = \rho(|A|)$ is satisfied [14]. If the parameters are allowed to contain intervals too, then the upper triangularity will be satisfied only for the nominal value for the interval $A(\Box \hat{X}(k))$. If for the interval matrix A: $\rho(A) < \rho(|A|) < 1$, then the Lohner's algorithm will produce an outer solution. And, if $\rho(A) < 1$ and $\rho(|A|) > 1$ the Lohner's algorithm will produce an unstable simulation/observation.

4.3 Neumaier's Algorithm [15]

Instead of using an interval hull of the set of possible estimated state $\hat{X}(k)$, Neumaier [15] proposes to use ellipsoids as enclosure sets and a new method for reducing the wrapping effect based on an interval ellipsoid arithmetic. In this paper, only the algorithm for the linear case will be presented. The extension for the non-linear case is a very simple task, since the non-linear system will be again linearised around the estimated trajectory [15].

An ellipsoid is the set of the form:

$$E(z, L, r) = \{z + L\xi \mid \xi \in \Re^n, \|\xi\| \le r, r > 0\} \tag{15}$$

where $z \in \Re^n$ is the centre, $L \in \Re^{n \times n}$ is the axis matrix and $r \in \Re$ is the radius.

The algorithm consists in propagating separately the center and the radius of the ellipsoid, being implicitly relative. Briefly it can be resumed as follows:

At each iteration, the radius of the new ellipsoid that enclose the uncertain domain with the relation is computed:

$$\tilde{r} \le \|\overline{L}^{-1} B\| r + \|\overline{L}^{-1} D\| q, \tag{16}$$

where: L is the ellipsoid shape, $B \approx mid(AL)$,

$$D \approx Diag\left(d_1 + d_1' r, \ldots, d_n + d_n' r\right), \tag{17}$$

$d \ge |Az + b - mid(Az + b)|$, A is the interval system matrix, and z represent the ellipsoid centre,

$$d' \ge |AL - mid(AL)|, \quad q \ge \|D^{-1} d\| + \|D^{-1} d'\| r. \tag{18}$$

The smallest box containing the ellipsoid $E(z, L, r)$ is

$$x := \square E(z, L, r) = z + [-r, r] \|L_{i\bullet}\|, \tag{19}$$

where $i\bullet$ represent the i-th row of the matrix L.

Using this algorithm, the parameters uncertainty can be managed without considering the extended system (including parameters as extra states) as in the case of Moore's and Lohner's algorithms.

The advantage of using ellipsoids instead of parallelepipeds is that the rotation of the state space of the interval system is implicitly in the case of ellipses propagation being not necessary to make additionally computations.

The disadvantage is that the algorithm for computing with ellipsoids is more complicated than computing with parallelepipeds, as in Moore and Lohner algorithms. Another drawback is that using the ellipsoids as enclosure sets there are some initial states, if the initial region of uncertainty is given by a box, that are not taken into account in case of taking the minimum volume ellipsoid fitting inside the box, obtaining then, a reduced space of possible states.

4.4 Kühn's Algorithm [10]

Kühn's algorithm is based on approximating the region of system states using zonotopes.

A zonotope Z of order m is the Minkowski sum

$$Z = \mathcal{P}^1 + \cdots + \mathcal{P}^m \tag{20}$$

of m parallelepipeds $\mathcal{P}^i$ (Figure 2). The order m is a measure for the geometrical complexity of the zonotopes. It can be chosen freely and is a performance parameter for the Kuhn's algorithm.

Given the zonotope Z_{k-1} enclosing the set of estimated states $\hat{X}(k-1)$ by system observer (3), then the set of estimated states $\hat{X}(k)$ is enclosed by the following zonotope

$$Z_k = \mathcal{R}(E_k + T_k Z_{k-1}) \tag{21}$$

where T_k are square matrices and E_k are intervals such that

$$f(Z_{k-1}) \subseteq E_k + T_k Z_{k-1} \tag{22}$$

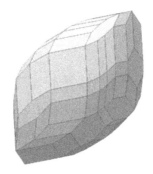

Fig. 2. A zonotope of order m=14

and the reduction operator $\mathcal{R}$ is defined in the following way: let $Z = \mathcal{P}^0 + \mathcal{P}^1 + \cdots + \mathcal{P}^m$ be a $m+1$ zonotope and $1 \le \ell \le m$ be the largest integer such that the following relation between diameters holds:

$$diam(\mathcal{P}^0 + \mathcal{P}^1 + \cdots + \mathcal{P}^{\ell-1}) \ge diam(\mathcal{P}^\ell) \tag{23}$$

or $\ell = 1$ otherwise, then:

$$\mathcal{R}(Z) := \Box\ (\mathcal{P}^0 + \mathcal{P}^1 + \cdots + \mathcal{P}^\ell) + \mathcal{P}^{\ell+1} + \cdots + \mathcal{P}^m \tag{24}$$

For the uncertainty in the parameters, the extended system (parameters as time-invariant states) must be used [19] as in the case of Lohner's and Moore's algorithms.

Kühn's algorithm can manage a uncertainty propagation better that the Lohner's and Neumaier's algorithms because it uses zonotopes for enclosing the uncertainty instead of using a naive box enclosure. However, if the system is non isotonic and non contractive, the zonotope that includes the family of zonotopes at each time instant still will include spurious states that can derive in an unstable simulation/observation, especially in the case of parameter uncertainty.

5 Approaches Using Real Trajectories

In this section, algorithms based on propagating real trajectories instead of regions, developed by Kolev [9] and Stancu [25], will be presented.

5.1 Kolev's Algorithm [9]

According to Kolev [9], the following approximate solution to the interval simulation problem that provide an inner solution can be obtained by determining the interval vector $[\breve{x}(k)] = [\breve{x}(k), \overline{x}(k)]$ by solving the following global optimisation problems:

$$\overline{x}(k) = max\, x(k, x_o, \theta, u)$$

and

$$\breve{x}(k) = min\, x(k, x_0, \theta, u)$$

subject to:

$$\theta \in V(\Theta)$$
$$x_o \in V(X_o) \tag{25}$$

where $V(\Theta)$ and $V(X_o)$ denotes the set of vertices of the uncertain parameters and initial states sets, respectively. This interval simulation algorithm is known as a *vertices algorithm*.

According to Nickel [16], the inner solution provided by the vertices algorithm coincides with the exact interval hull of the solution set for some systems, those without the wrapping effect that verify that their state function is *isotonic* with respect to all state variables [4]. Moreover, for such systems, region based approaches and trajectory based approaches will provide the same results.

5.2 Optimisation Algorithm [25]

Finally, another algorithm for simulating/observing an interval linear system coming from the fault detection community was proposed in [25]. The algorithm is an extension of Puig's algorithm [23] for the non-linear case.

This algorithm is based on a linearisation of the state equations about the current state estimate according to (9), as in the Extended Kalman Filter (EKF), combined with an optimisation of the possible trajectories from the initial state to avoid the wrapping effect and parameter time-invariance problems. The idea of using linearisation to deal with the problem of interval observers has also been proposed by Shamma [24] and Calafiore [3]. Linearisation is required in order to design a stable observer, since linearised observer presented in (11) is linear parameter varying (LPV) where the scheduling variable is the central estimate $\hat{x}_c(k)$. Then, a stable observer for such an LPV observer can be obtained using LMI techniques [5].

Once the linearised observer is introduced, a similar optimisation based algorithm as is proposed in Puig [23] will be applied.

Algorithm 2. Interval observer based on optimisation

Let $y = \{ y(0), y(1), y(2),...y(k-1) \}$ be a measurement trajectory of system (1) and assuming that uncertainty on initial state is $x(0) \in X_0$:

- at each time step compute $\Box \hat{X}(k) = \left[\hat{\underline{x}}(k), \overline{\hat{x}}(k) \right]$, solving the following optimisation problem for each component of $x(k)$ to determine $\hat{\underline{x}}(k)$:

$\hat{\underline{x}}_i(k) = min\ (\ globally\ )\ \hat{x}_i(k)$

subject to :

$\hat{x}(k) = g_0(\ \hat{x}_c(k-1), u_0(k-1), \theta_c\ )$
$+ A(\ \hat{x}(k-1), \theta\ )(\ \hat{x}(k-1) - \hat{x}_c(k-1))$

...   (26)

$\hat{x}(1) = g_0(\ \hat{x}_c(0), u_0(0), \theta_c\ )$
$+ A(\ \hat{x}(0), \theta\ )(\ \hat{x}(0) - \hat{x}_c(0))$

$\hat{x}(j) \in \left[\hat{\underline{x}}(j), \overline{\hat{x}}(j) \right]$ for $j = 0,...,k$

$\theta \in \left[\underline{\theta}, \overline{\theta} \right]$

where: $g_0(\ \hat{x}(k), u_0(k), \theta)$ is   the   state   space   observer   function   and $u_o(k) = \left[u(k) \quad y(k) \right]^t$ is the observer input

- and solving again the previous optimisation problems substituting min by max to determine $\overline{\hat{x}}(k)$.

The previous algorithm guarantees that $\Box \hat{X}(k+1)$ includes the real uncertainty region since it is implicitly applied the **mean-value theorem**:

$$\Box \hat{X}(k+1) \in g_0(\ \hat{x}_c(k), u_0(k)) \\ + A(\Box \hat{X}(k), \theta)(\Box \hat{X}(k) - \hat{x}_c(k)) \tag{27}$$

One of the main drawbacks of this approach is the high computational complexity of the optimisation algorithm since at each iteration an additional restriction is added. So, the amount of computation needed is increasing with time being impossible to operate over a large time interval. Then, some kind of approximation should be introduced to make the approach more tractable. The length increase problem in the previous approach can be solved if the observer (5) is asymptotically stable.

Along a particular estimate trajectory $\hat{x}_e(k)$ and for a given $\theta \in \Theta$ using (11) again to approximate the non-linear observer, the following linear parameter varying system can be introduced

$$\hat{x}(k) = A(\hat{x}_e(k-1), \theta)\hat{x}(k-1) + B(\hat{x}_e(k-1), \theta)u_n(k-1) \tag{28}$$

with:

$$\tilde{A} = A(\hat{x}_e(k-1),\theta), \tilde{B} = [I \quad A(\hat{x}_e(k-1),\theta)] \text{ and}$$
$$u_n(k) = [g_o(\hat{x}_e(k),u_o(k),\theta) \quad \hat{x}_e(k-1)]^t.$$

Substituting recursively equation (28) in the objective function of (26) the following objective function can be obtained

$$\hat{x}(k) = \Phi(k,0,\theta)\hat{x}(0) + \sum_{j=0}^{k-1} \Phi(k,j,\theta)B(\hat{x}_e(j),\theta)u_n(j) \tag{29}$$

where:

$$\Phi(k,j,\theta) = \prod_{p=j}^{k-1} A(\hat{x}_e(p),\theta) \tag{30}$$

what allows to reformulate the optimisation problem (26) as it was done in [21] in the case of time-invariant linear interval observation algorithm, taking as estimated trajectory the central estimate $\hat{x}_c(k)$:

$$\underline{\hat{x}}(k) = \min \left[\begin{array}{c} \Phi(k,j,\theta)\hat{x}(0) \\ + \sum_{j=0}^{k-1} \Phi(k,j,\theta)B(\hat{x}_c(k),\theta)u_n(j) \end{array} \right]$$

$$subject\ to:$$

$$\Phi(k,\mathbf{0},\theta) = \prod_{p=j}^{k-1} A(\hat{x}_c(p),\theta)$$

$$\hat{x}(0) \in \left[\underline{\hat{x}}(0), \overline{\hat{x}}(0)\right] \tag{31}$$

$$\theta \in \left[\underline{\theta}, \overline{\theta}\right]$$

If the interval observer (5) were a linear time-invariant (LTI) system, stable for all $\theta \in \Theta$, then it would exist a temporal horizon L such that:

$$\left\|A^L(\theta)\right\|_\infty < 1 \tag{32}$$

for all $\theta \in \Theta$ that can be determined using results presented in [23]. Then, the interval observation produced by (26) using this temporal horizon will avoid the instabilisation effects produced by the wrapping effect. However, since the linearised interval observer (28) is a linear parameter varying system, in order to apply the same idea, the following assumption is proposed:

Assumption 1:

Condition (32) for a stable linear parameter varying (LPV) observer will imply:

$$\left\|A(x_{k-L},\theta)A(x_{k-L+1},\theta)...A(x_k,\theta)\right\|_\infty < 1 \tag{33}$$

for all $\theta \in \Theta$.

Then using such approximation, the *Algorithm 2* can be formulated in a more tractable way since, for any time k, the optimisation problem (26) will be approximated using a sliding window, starting at time k-L and ending at k, according to

$$\underline{\hat{x}}(k) = min \begin{bmatrix} \boldsymbol{\Phi}(k,j,\boldsymbol{\theta})\hat{x}(k-L) + \\ \displaystyle\sum_{j=k-L}^{k-1} \boldsymbol{\Phi}(k,j,\boldsymbol{\theta})\boldsymbol{B}(\boldsymbol{\theta},\hat{\boldsymbol{x}}_c(p))\boldsymbol{u}_n(j) \end{bmatrix}$$

subject to :

$$\boldsymbol{\Phi}(k,j,\boldsymbol{\theta}) = \prod_{p=j}^{k-1} \boldsymbol{A}(\hat{\boldsymbol{x}}_c(p),\boldsymbol{\theta}) \qquad\qquad (34)$$

$$\hat{x}(k-L) \in \left[\underline{\hat{x}}(k-L), \overline{\hat{x}}(k-L)\right]$$

$$\boldsymbol{\theta} \in \left[\underline{\boldsymbol{\theta}}, \overline{\boldsymbol{\theta}}\right]$$

where L is the length of this window that satisfies the relation (33).

The linearised observer (21), proposed in this section to approximate *Algorithm 1*, allows to solve two stabilisation problems:

- the first consists in designing of a stable observer (K) using LPV observer theory [5],
- and the second consists in determining a time window (L) using *Assumption 1* such that avoids the instability produced by the wrapping effect and preserve uncertain parameter time invariance.

If the interval observer satisfies the isotony property, i.e. the variation of the state function (4) respect all the states is positive, only the first stabilisation problem should be considered since the wrapping effect is not present [24].

6 Comparison of the Algorithms

This section is dedicated to test all the algorithms presented in this paper. Two benchmark problems will be used. The first example is based on an interval system used as a case study by Neumaier [15] while the second one is a complex non-linear system proposed in an European project DAMADICS [1] as a fault detection benchmark.

In order to show the effectiveness in propagating state uncertainty previous algorithms will be tested when applied to solve the interval observation problem in the hardest conditions, i.e., when observer gain L is equal to zero (interval simulation). It is known that selecting the observer gain adequately, the resulting observer could satisfy the condition of isotony [7] and all algorithms will provide the same results. It is an open problem to be addressed in further papers the design of the observer gain in order to satisfy such condition and at the same time the fault detection and stability requirements, among others.

6.1 Test Example 1

First, an example proposed in Neumaier [15] will be used in order to compare algorithm's performance:

$$x_1(k) = px_1(k-1) + px_2(k-1) + b_1$$
$$x_2(k) = -px_1(k-1) + px_2(k-1) + b_2 \qquad (35)$$
$$p(k) = p(k-1)$$

with uncertain initial conditions: $x_1(0) \in [-1,1]$, $x_2(0) \in [-1,1]$

and parameters: $b_1 \in \left[-10^{-12}, 10^{12}\right]$, $b_2 \in \left[-10^{-12}, 10^{12}\right]$ and $p \in [0.4, 0.5]$.

Since the interval system matrix $A = \begin{bmatrix} p & p \\ -p & p \end{bmatrix}$ do not fulfils the condition of

isotony, the system suffers from the wrapping effect .

For the given interval on the system parameter p, the system will be at limit contractive, i.e. $\|A(p)\|_\infty = 1$. In this case, the algorithms which use the region propagation[2], except the naive approach based on the absolute Moore's algorithm (*Section 5.1*), avoid the instability because of the wrapping effect, but only provide an outer solution with a certain degree of conservatism depending on the kind of the geometry used to approximate the real uncertainty region. Neumaier's (*Section 5.3*) and Kühn's (*Section 5.4*) algorithms provide a better approximation, since the use the ellipsoids and zonotopes (of order $m=5$), respectively, than Lohner's (*Section 5.2*) algorithm which use parallelepipeds. In this example, Kolev (*Section 6.1*) and optimisation (*Section 6.2*) algorithms provide the same results.

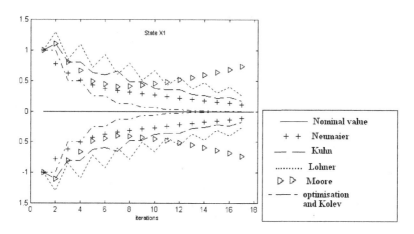

Fig. 3. Comparison between the algorithms in the case $\|A(p)\|_\infty = 1$

---

[2] The algorithms from Section 4 were implemented using INTLAB V3.1 package and MATLAB 6.5, except Kühn's algorithm which has been adapted to this example from his JAVA implementation.

If the parameter uncertainty is changed such that $p \in [0.2, 0.3]$, then the interval system will be contractive, i.e. the infinity norm of the system matrix changes to $\|A(p)\|_\infty \leq 0.6$. In this case the region based methods will compute an outer approximation of the real region. Kolev's algorithm (*Section 6.1*) will provide a better inner solution, and optimisation algorithm will provide an outer solution not very far from the exact one (Figure 4).

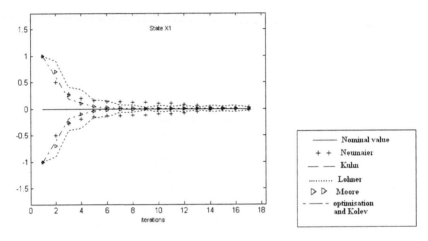

Fig. 4. Comparison between the algorithms in the case $\|A(p)\|_\infty \leq 0.6$

The conservatism of the solution computed by region propagation approaches depends of the region used by each algorithm to approximate the region of possible states and on the uncertainty propagation strategy. Better results are obtained with Kuhn (*Section 5.4*) and Neumaier (*Section 5.3*) algorithms since they use zonotopes and ellipsoids that provide a better approximation that a parallelepiped or a box. In case of contractive system, all region based methods give stable interval simulation/observation. If the parameter uncertainty is increased changing the interval on parameter p to be $p \in [0.4, 0.7]$, the infinity norm of the system matrix changes to $\|A(p)\|_\infty \leq 1.4$.

Then, the wrapping effect will increase at each time step providing an unstable simulation in the case of Moore's (*Section 5.1*), Lohner's (*Section 5.2*), Neumaier's (*Section 5.3*) and Kuhn's (*Section 5.4*) algorithms (see *Table 1*). Comparing trajectory based algorithms with the region based using *Table 1*, it can be observed that the first avoid the wrapping effect. Kolev's (*Section 6.1*) algorithm provides an inner solution, while optimisation (*Section 6.3*) algorithm provides the exact solution with a given precision.

Table 1. Comparison between the algorithms for test example 1

Time (in s)	10	20	30	40
Optimisation	1.6588	0.2683	0.1097	-0.0530
Kolev	1.5873	0.2624	0.0342	-0.0646
Neumaier	10.1109	50.6687	267.7928	1.3726e+03
Kuhn	3.8515	11.5277	52.9601	200.7197
Lohner	390.3358	4.0736e+10	3.5576e+27	4.5002e+53
Moore	4.4140e+06	∞	∞	∞

(the results present upper bound for the estimated interval corresponding to state variable x_1)

6.2 Test Example 2

The second test example 2 deals with an industrial smart actuator consisting of a flow servo-valve driven by a smart positioner, proposed as a fault detection benchmark in the European DAMADICS project. The smart actuator consists of a control valve, a pneumatic servomotor and a smart positioner [1]. In this test example, we will focus on the pneumatic servomotor and the electro-pneumatic transducer.

The non-linear interval model is obtained using interval model identification techniques in fault-free scenario, as those proposed in Ploix [17]. The identified non-linear interval model will be:

$$x_1(k+1) = x_1(k) + \Delta\left(\theta_1 x_2(k) + \theta_2\right)$$
$$x_2(k+1) = x_2(k) + \Delta\left(-a_{21}x_1(k) - a_{22}x_2(k) + a_{23}x_3(k) + c_2\right)$$
$$x_3(k+1) = x_3(k) + \Delta\left(-a_{32}x_2(k)\frac{x_3(k)+P_a}{V_0 + A_e x_1(k)} + a_{34}\frac{1}{x_4(k)}\frac{x_4(k)-x_4(k-1)}{\Delta}\left(x_3(k)+P_a\right)\right)$$
$$x_4(k+1) = x_4(k) + \Delta\left(k_1 CVP\sqrt{P_z - x_3(k)}f_{p1} + k_1 CVP\sqrt{x_3(k)}f_{p2}\right)$$

where Δ is the discretisation step size and the uncertain parameters are: $\theta_1 \in \left[\underline{\theta_1}, \overline{\theta_1}\right]$, $\theta_2 \in \left[\underline{\theta_2}, \overline{\theta_2}\right]$, $a_{21} \in \left[\underline{a_{21}}, \overline{a_{21}}\right]$, and $a_{22} \in \left[\underline{a_{22}}, \overline{a_{22}}\right]$.

The system suffers again from the wrapping effect because it does not fulfil the property of isotony. As we will see in the following, Moore's (*Section 5.1*), Lohner's (*Section 5.2*), Neumaier's (*Section 5.3*) and Kuhn's (*Section 5.4*) algorithms will provide an unstable interval simulation. On the other hand, interval simulation obtained with the optimisation algorithm (Section 6.2) for 40000 iterations are presented in Figure 5.

The optimisation algorithm (*Section 6.2*) provides the exact solution (using an infinite time horizon) and Kolev's (*Section 6.1*) algorithm provides an inner solution being very close to the exact (*Table 1*). As it can be observed from *Table 1*(upper bound for the estimated interval corresponding to state x_1), Moore's (*Section 5.1*),

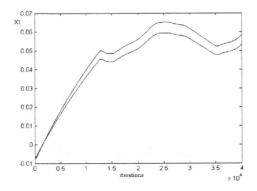

Fig. 5. The envelopes for the pneumatic servomotor

Lohner's (*Section 5.2*) and Neumaier's (*Section 5.3*) algorithms will inflate very quickly the system state interval. For example in the case of Moore's algorithm after 208 iterations $P_z - x_3(k)$ will be negative being impossible to compute a real value for $\sqrt{P_z - x_3(k)}$. However, we can see that Kuhn's (*Section 5.4*) algorithm manage better the interval system since the use of a more complex uncertainty approximating regions (zonotopes). The algorithm fails after 475 iterations.

Table 2. Comparison between the algorithms for test example 2

Time (in s)	200	300	400	25000	30000
Optimisation	- 0.0067	- 0.0062	- 0.0057	0.0651	0.0619
Kolev	- 0.0067	- 0.0062	- 0.0057	0.0642	0.0589
Neumaier	- 0.0049	- 0.0946	∞	∞	∞
Kuhn	- 0.0041	- 0.0009	0.0080	∞	∞
Lohner	0.1143	∞	∞	∞	∞
Moore	0.1234	∞	∞	∞	∞

(the results present upper bound for the estimated interval corresponding to state variable x_1)

6.3 Final Comments

In fault detection applications the real-time operation is needed. Moore's, Lohner's, Neumaier's and Kuhn's algorithms compute the envelopes very efficiently using one step ahead iteration suitable to be used in real-time but only for a particular case of systems, i.e. isotonic systems, provides the exact solution. When the system is not isotonic, and the system parameters are in intervals, in general these algorithms do not avoid the wrapping effect. However for isotonic systems Kolev algorithm provides the exact solution.

As we have seen in the *Section 7.1* of this paper, the wrapping effect using region based approaches can be avoided for small intervals over model parameters. Also, in the *Section 7.2* we have seen that for the interval $p \in \left[\dfrac{2}{10}, \dfrac{3}{10} \right]$ the system proposed as example is contractive. On the other hand, optimisation algorithm (*Section 6.2*) avoids the wrapping effect also for non isotonic system, but an exponential computational time is needed. For the first example presented above for 20 iterations the computational time was greater than 10 hours (Pentium 4, 2.4GHz). However, the Kolev's algorithm (inner solution) provides a good approximation for the exact solution for the non isotonic systems. Kolev's (*Section 6.1*) algorithm performs in real-time since propagations of a limited number of trajectories (corresponding to vertices of parameter region).

7 Conclusions

This paper is a first try to benchmarking several existing for interval simulation/observation algorithms developed in the different research areas applied to nonlinear systems with uncertain parameters and its comparison with a new algorithm based on optimisation (*Section 6.2*). Region algorithms are based on propagating the uncertainty region for system states using one step-ahead recursion. The main problem of region based interval observation/simulation is the wrapping effect. This problem prevents the use of the naive absolute Moore's algorithm when it is present. In this case, more sophisticated approaches should be used as: relative Moore's, Lohner's, Neumaier's and Kühn's algorithms. However, these algorithms fail when the system do not fulfil the isotonoy property because of parametric uncertainty. In this case, the region that include all possible states at each time instant will contain spurious states that will inflate the region and in many cases the interval simulation/observation will be unstable, as it is presented in the proposed test examples. These results reinforce the use of algorithms based on the propagation of real trajectories instead of regions as in the algorithms presented in the *Section 6*: Kolev's (*Section 6.1*), optimisation (*Section 6.2*) algorithms. However, since Kolev's algorithm provide an inner solution, and since the optimisation algorithm is more time consuming, these algorithms should be improved in order to be applied in fault detection applications where real-time operation and completeness of the simulation is needed. On possible improvement is presented in Stancu [26] where Kolev's algorithm is combined with a complementary test based on constraint satisfaction algorithms.

After analysing the results presented in this paper, we can conclude that although region based approaches look appealing because their lower complexity compared with trajectory based approaches in many cases they can derive in unstable observations because of the wrapping effect. This seems to reinforce the use of trajectory based approaches, but still in this case the computational complexity limits their applicability in fault detection where real-time computations are required. Reached this point, the need to design the observer gain such that the isotony condition [4][7] be satisfied seems a possible solution. In this case region based approaches will not suffer from the wrapping effect and will provide the same results as trajectory based approaches. This should be further investigated since not only the isotony condition should be satisfied when designing an observer for fault detection since there are other requirements to be satisfied.

Acknowledgements

This paper is supported by CICYT (DPI2002-03500), by Research Commission of the "Generalitat de Catalunya" (group SAC ref.2001/SGR/00236) and by DAMADICS FP5 European Research Training Network (ref. ECC-TRN1-1999-00392).

References

1. Bartys, M.Z. "Specification of actuators intended to use for benchmark definition". http://diag.mchtr.pw.edu.pl/damadics/. 2002.
2. Chen J. and R.J. Patton. "Robust Model-Based Fault Diagnosis for Dynamic Systems". Kluwer Academic Publishers. 1999.
3. Calafiore, G. "A Set-valued Non-linear Filter for Robust Localization". In Proceeding of European Control Conference 2001 (ECC'01). Porto. Portugal. 2001.
4. Cugueró, P., Puig, V., Saludes, J., Escobet, T. "A Class of Uncertain Linear Interval Models for which a Set Based Robust Simulation can be Reduced to Few Pointwise Simulations". In Proceedings of Conference on Decision and Control 2002 (CDC'02). Las Vegas. USA. 2002.
5. Daafouz, J., Bara, G. I., Kratz, F., Ragot, J., "State Observers for Discrete-Time LPV Systems", 39th IEEE Conference on Control Decision and Control, 2001.
6. ElGhaoui, L., Calafiore, G. "Worst-Case Simulation of Uncertain Systems". Robustness in Identification and Control. A. Garulli, A. Tesi & A. Vicino Eds. Springer.
7. Gouzé, J.L., Rapaport, A, Hadj-Sadok, M.Z. "Interval observers for uncertain biological systems". Ecological Modelling, No. 133, pp. 45-56, 2000.
8. Horak, D.T. "Failure detection in dynamic systems with modelling errors" J. Guidance, Control and Dynamics, 11 (6), 508-516.
9. Kolev, L.V. "Interval Methods for Circuit Analysis". Singapore. World Scientific. 1993.
10. Kühn, W. "Rigorously computed orbits of dynamical systems vithout the wrapping effect". Computing, 61(1), pp. 47-67. 1998.
11. Lohner, R.J. "Enclosing the Solution of Ordinary Initial and Boundary Value Problems", in Kaucher, E., Kulisch, U. & Ullrich, Ch. (eds.): Computerarithmetic: Scientific Computation and Programming Languages. B.G. Teubner, pages 255-286. Stuttgart. 1987.
12. Moore, R.E. "Interval analysis". Prentice Hall. 1966.
13. Moore, R.E. "Methods and applications of interval analysis". SIAM. Philadelphia. 1979.
14. Nedialkov, N.S., Jackson, K.R. "A New Perspective of the Wrapping Effect in Interval Methods for Initial Value Problems for Ordinary Differential Equations". In Kulisch, Lohner and Facius Eds. "Perspectives on Enclosure Methods", pp. 219-264. Springer-Verlag. 2001.
15. Neumaier, A. "The wrapping effect, ellipsoid arithmetic, stability and confidence regions". Computing Supplementum, 9, pp. 175-190. 1993.
16. Nickel, K. "How to Fight the Wrapping Effect". In K. Nickel, editor, Interval Analysis 1985, Lecture Notes in Computer Science, 212, pp. 121-132. Springer. 1985.
17. Ploix, S., Adrot, O., Ragot, J. "Parameter Uncertainty Computation in Static Linear Models". 38th IEEE Conference on Decision and Control. Phoenix. Arizona. USA.

18. Puig, V., Saludes, J., Quevedo, J. "A new algorithm for adaptive threshold generation in robust fault detection based on a sliding window and global optimisation". In Proceedings of European Control Conference 1999, ECC'99. Germany, September. 1999.

19. Puig, V., Cugueró, P., Quevedo, J. "Worst-Case Estimation and Simulation of Uncertain Discrete-Time Systems using Zonotopes". In *Proceedings of European Control Conference 2001, (ECC'01)*. Portugal. September. 2001.

20. Puig, V., Quevedo, J., Escobet, T., De las Heras, S. "Robust Fault Detection Approaches using Interval Models". IFAC World Congress (b'02). Barcelona. Spain. 2002.

21. Puig, V., Cugueró P., J., Quevedo, J., Escobet, T. "Time-invariant approach to worst-case simulation and observation of discrete-time uncertain systems". In Proceedings of Conference on Decision and Control 2002 (CDC'02). Las Vegas. USA. 2002.

22. Puig, V., Quevedo, J., Escobet, T., Stancu, A. "Passive Robust Fault Detection using Linear Interval Observers". IFAC Safe Process, 2003. Washington. USA.

23. Puig, V., Saludes, J., Quevedo, J. "Worst-Case Simulation of Discrete Linear Time-Invariant Dynamic Systems", Reliable Computing 9(4): 251-290, August. 2003.

24. Shamma, J.S. "Approximate Set-Value Observer for Nonlinear Systems". IEEE Transactions on Automatic Control, Vol. 42, No 5. 1997.

25. Stancu, A., Puig, V., Quevedo, J., Patton R. J. "Passive Robust Fault Detection using Non-Linear Interval Observers: Application to the DAMADICS Benchmark Problem". IFAC Safe Process, 2003. Washington. USA.

26. Stancu, A., Puig, V., Quevedo, J. "Gas Turbine Model-Based Robust Fault Detection Using a Forward – Backward Test", 2nd International Workshop on Global Constrained Optimization and Constraint Satisfaction (Cocos '03), November 2003, Lausanne, Switzerland.

Author Index

Lecture Notes in Computer Science

For information about Vols. 1–3402

please contact your bookseller or Springer